A Laboratory Guide for Identification of Plant Pathogenic Fungi

NIPA® GENX ELECTRONIC RESOURCES & SOLUTIONS P. LTD.
New Delhi-110 034

About the Authors

Professor S.G. Borkar is an alumnus of the Indian Agricultural Research Institute, New Delhi; INRA Angers, France; and International University, Washington, USA. He has more than 38 years of professional experience in the subject of Plant Pathology at three Agricultural universities in two states. He has more than 135 research articles published in national and foreign journals, including 10 books on his subject in India and the USA, and six patents. He has mentored 30 students for their Masters and Doctorate degrees in Plant Pathology.

The co-authors of the book are presently working as Assistant Professors of Plant Pathology at different Agricultural Universities. Ajayasree T.S. is at Kerala Agriculture University, P. Srilatha is at Professor Jayashankar Telangana Agriculture University, and Jayashree Bhattacharjee is at Cultyvate with wide experience in teaching the subject of fungal pathology.

A Laboratory Guide for Identification of Plant Pathogenic Fungi

S.G. Borkar
Former Head
Department of Plant Pathology
Mahatma Phule Agricultural University
Rahuri, Maharashtra

Ajayasree T. S.
Department of Plant Pathology
Kerala Agricultural University
Thrissur, Kerala

P. Srilatha
Professor Jayashankar Telangana State Agriculture University
Hyderabad, Telangana

Jayashree Bhattacharjee
Implementation Lead Agronomy
Cultyvate, Bangalore, Karnataka

NIPA® GENX ELECTRONIC RESOURCES & SOLUTIONS P. LTD.
New Delhi-110 034

NIPA. GENX ELECTRONIC RESOURCES & SOLUTIONS P. LTD.

101,103, Vikas Surya Plaza, CU Block
L.S.C.Market, Pitam Pura, New Delhi-110 034
Ph : +91 11 27341616, 27341717, 27341718
E-mail:newindiapublishingagency@gmail.com
www: www.nipabooks.com

For customer assistance, please contact
Phone: + 91-11-27 34 17 17
Fax: + 91-11- 27 34 16 16

Print ISBN: 978-93-58876-54-3

ebook ISBN: 978-93-58871-69-2

Composed and Designed by NIPA

Acknowledgments

We are grateful to all our colleague professors and fungal pathologists across the world whose photographs are used for this book and are in the public domain. In the absence of the availability of their names in the public domain, the courtesy of their names in the photographs is not mentioned; however, these are of those individual scientists who are anonymous to us.

Our thanks are to all our teachers in the subject of mycology and fungal plant pathogens and our graduate, post-graduate, and doctorate students, in India and abroad, whose demands and inspirations made us write such an important title.

Prof. Suresh Borkar et al.

Preface

Humans encountered plant diseases during their nomadic lives and the early plant pathogens caused these diseases were fungal plant pathogens. These fungal pathogens attack almost all crops and plants by attacking different plant parts like root, trunk, stem, twigs, leaves, buds, inflorescence, flowers, fruits, and seed to cause different types of symptoms at the site of infection. The diseases are named on the basis of symptoms developed by fungal infection like damping off, seedling blight, wilt, root rot, collar rot, gummosis, twig blight, die-back, leaf spot, leaf blight, fire blight, rust, downy mildew, powdery mildew, smut, bunt, malformation, anthracnose, fruit rot, scab, pustules, seed rot, etc. These diseases may take sporadic, endemic, or epidemic forms to cause losses in crop production and crop produce, which directly affect the availability of food or quality food for humans and animals. Around 20,000 species of fungal pathogens attack different plants around the world to make them diseased, and furthermore, different fungal pathogens can cause the same disease and symptoms. Therefore, the fungal pathogen involved as a cause of the disease is most important to be identified for their control measures as different chemical molecules are effective for different fungal pathogens.

The fungal pathogen involved as a cause of the disease can be directly identified from the diseased sample if the microscopic fungal structures necessary for fungal genus or species identification are available in the infected sample. Otherwise, the fungal pathogen has to be isolated in pure form, confirmed for pathogenicity, and then identified either based on morphological characteristics or molecular characterization.

The present book, "A Laboratory Guide for Identification of Plant Pathogenic Fungi," summarizes all the basic and detailed information required in the identification of fungal plant pathogens. This will help the teachers and students of Botany, Plant Pathology, Forestry, Vegetable Science, Silviculture, Fodder and Pastures Science, Life-Sciences, and all those who study fungal plant diseases.

Prof. Suresh Borkar et al.

Contents

1

The Fungi
A Dreaded Pathogen for Crop Plants

Fungi have been known to cause plant diseases in crops and plants of economic importance to humans since time immemorial. Ancient agriculture witnessed these plants' sufferings but thought it was due to the curse of God on humans. The curiosity to know the causes of these crops and plant suffering developed the science of plant pathology and enlightened our understanding of the causes of these diseases. Among different plant pathogens, fungi occupy the prime position to cause plant diseases. They attack almost all the crops and plants to make them diseased. Around 20,000 species of fungal pathogens attack different plants around the world to make them diseased (USEPA 2005). These fungi attack almost all the parts of the plants i.e. root, trunk, stem, twigs, leaves, buds, inflorescence, flowers, fruits, and seeds to cause different types of symptoms at the site of infection and the diseases are named based on symptoms these fungi cause like damping off, seedling blight, wilt, root rot, collar rot, gummosis, twig blight, die-back, leaf spot, leaf blight, fire blight, rust, downy mildew, powdery mildew, smut, bunt, malformation, anthracnose, fruit rot, scab, pustules, seed rot, etc. These diseases may be sporadic, endemic, or in epidemic form.

The damages due to these fungal pathogens and their diseases on different crops are enormous, causing economic losses, sometimes human starvation, human migrations, and political instability. The known examples of such diseases are the late blight of potato epidemic in Ireland and southern corn blight in the USA, stem rust of cereals around the world, bitter rot of apples, chestnut blight epidemic in the USA, downy mildew epidemic of grapes in Europe, Asian soybean rust, wheat blast and blotch in the EU, banana wilt in African countries, cassava root rot, sunflower downy mildew, Ash dieback in the United Kingdom, Bengal famine due to rice disease in British India, etc.

Out of the 220 billion dollars in losses due to plant diseases around the world in a year, at least 70–80% of losses are due to fungal diseases means that the annual losses caused due to fungal plant diseases range up to 154 billion US dollars. In terms of food losses, the fungal infections presently destroy at least

125 million tonnes of the top five food crops—rice, wheat, maize, potatoes, and soybeans each year. With the management of fungal crop plant diseases, the monetary losses of at least 154 billion US$ can be saved and this amount can be used for the well-being of the food-deprived or hungry people in the world. As per FAO estimates of 2020 in comparison with 2019, 46 million more people in Africa (out of 282 million), almost 57 million more in Asia (out of 418 million), and about 14 million more (out of 60 million) in Latin America and the Caribbean were affected by hunger in 2020. As per estimates of different agencies, the cost of ending hunger in the US is $25 billion (Joel Berg, CEO of Hunger Free America, 2021), and wiping out hunger in Africa could cost just $5 billion (Feike Sijbesma, 2019) or one can feed 2.2 billion malnourished children on earth for a year (GlobalGiving Team 2021) with the amount we lose by fungal plant diseases. Therefore, the management of fungal plant diseases and the pathogenic fungi as micro-organisms is important and this is possible only if we can identify them correctly since the fungicides used for the management of different fungal species or the diseases caused by them, are molecules of different chemistry. Thus, accurate identification of plant pathogenic fungi is important to save our crops and plant treasures from their invasion.

1.1. What are Fungi

These are micro-organisms having filamentous thread-like structures known as hyphae, numerous hyphae interwoven together to form mycelium, which may bear their fruiting structures and spores (seeds). These are generally microscopic but can be visualized in the form of fungal colonies on their growth substrate and medium or their infection site on plant parts.

1.2. How to Recognize the Fungal Infection on/in Plants

The fungal-infected plant exhibits the typical symptoms produced by the concerned fungus, i.e. the transformation of grains into fungal spore sori in the case of smut and bunt disease, the appearance of rust pustules on infected leaf surfaces, powdery fungal growth in powdery mildew infection, downy fungal growth underside of the leaf surface in downy mildew, spot or blighted portions on infected leaves for the leaf spot/blight pathogen, or wilting of the plant in the case of wilt diseases, etc.

In certain diseases, the presence of fungal structures like fruiting bodies of the fungus, or growth of the fungus on infected plant parts may be seen while in other diseases these are present inside the host tissues. Observation of the infected portion of the host under the simple or binocular microscope certainly reveals the presence of fungi.

1.3. How to Differentiate Fungal Colonies on Culture Media from Other Microorganisms

The fungal colony on the culture media exhibits typical flappy, fussy mycelial growth of the fungus while the growth of bacterial colonies is non-flappy and non-fussy. Bacterial colonies may be pinhead, circular, irregular, raised or suppressed, white or colored (Fig. 1.1).

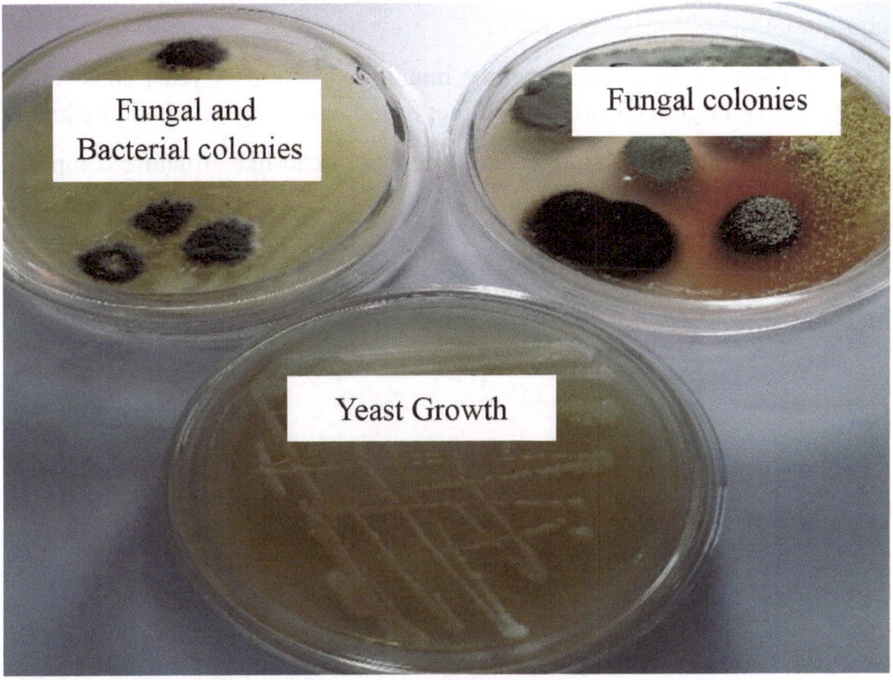

Fig. 1.1: Colonies of micro-organisms on culture media.

The initial growth of all fungal colonies is white for a day but changes color later on, depending on their genera and species.

No fungal colony in the growth media emits any type of odor whereas most contaminant bacterial colonies emit a foul odor.

The colonies of actinomycetes look like fungal colonies, but these appear suppressed and interwoven in the media.

The yeast colonies on growth media as contaminants also appear as non-flappy and non-fussy and appear like the raised bacterial colonies.

1.4. How to Differentiate Fungi from Other Microbes Under a Microscope

Under a binocular microscope, the stained sample of fungus exhibits a thread-like fungal structure which may be septate or non-septate and may have hyaline or colored spores.

The stain sample of bacteria does not show such a mycelial structure. The bacteria may be rod-shaped or cocci-shaped and numerous, may be motile or non-motile.

The stain sample of yeast shows the budding structure of yeast cells. These are generally single-cell structures either round or oblong with budding knobs.

The stain sample of Actinomycetes exhibits a thread-like structure composed of several individual cells (Fig. 1.2).

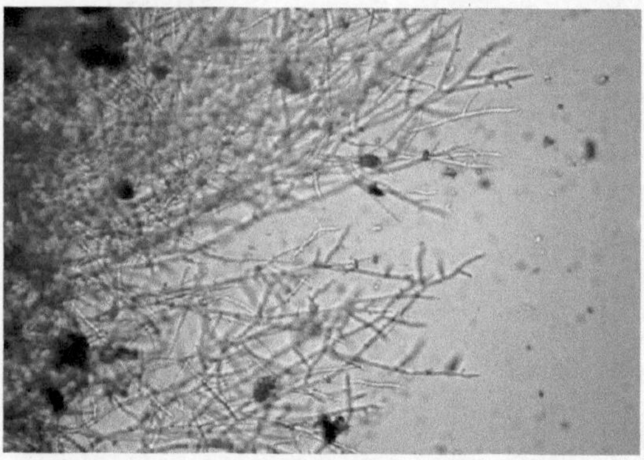

Fig. 1.2: Fungal structure in binocular microscope.

2

Microscopy of Fungal Specimens

For visualization and identification of fungal specimens into their genus and species, the staining of the fungal specimens and their microscopy are necessary. The simple binocular microscope can be used for microscopy of the fungal specimens.

2.1. Stains to be Used in Microscopy

1. Methylene blue stain
2. Lacto phenol cotton blue stain
3. Crystal violet stain
4. Safranin

For preparation of these stains please refer to Appendices.

2.2. Microscopy of Disease Specimens

For microscopy of collected disease samples follow the following steps:

- Wash the collected diseased samples gently with water (root samples should be washed in tap water to remove the soil). Leaf, and twig samples should be cleaned with a moistened cotton swab except for powdery and downy mildew disease to remove the dirt present on the sample.
- Cut an appropriate piece of the disease sample and place it on a clean glass slide in a drop of distilled sterile water. Dissect with needles or cut the disease tissues with a sharp blade into small pieces. Take care to have sufficient water in the dissected sample.
- Add 1–2 drops of appropriate stain on the dissected sample and leave for 2–3 min.
- Observe the sample under the microscope to detect the presence or association of fungal structures, fungal spores, or fruiting bodies with the diseased specimen.
- Based on the fungal structures, fungal spores, or fruiting bodies the fungus is identified at the genus level.

In a Methylene blue stain, the fungal structure acquires the stain and appears blue; while in a Safranin stain, the host tissues take safranin color while the fungal structures remain colorless.

2.3. Microscopy of Fungal Specimens

For microscopy of fungal specimens follow the following steps:

- Take a clean glass slide and add a drop of sterile water at the center of the slide. Add a loop of fungal culture to the drop of water.
- Add methylene blue or lacto phenol stain to this preparation and cover with a microscopic cover slip. Remove the air bubbles, if any, present in the preparation.
- Observe the fungal sample in the binocular microscope for the mycelial structures, fungal fruiting bodies, spore-bearing structures, and fungal spores.
- Measure the fungal structures with the help of a calibrated ocular micrometer, and note down properly the length and breadth of each structure.
- Take a microphotograph of the specimen with the help of the appropriate camera.
- Based on the structures and size of sporophores, spores, or fruiting bodies, the fungus is identified up to the genus or species level.

3

Glossary
Used in the Identification of Fungi

3.1. Terminology Used in the Description of the Fungal Pathogen

3.1.1. The Mycelium

The hypha or mass of hyphae that make up the body of a fungus.

3.1.1.1. Coenocytic or aseptate mycelium

Mycelium in which the hyphae lack septa is called coenocytic or aseptate mycelium. The aseptate hyphae have nuclei scattered in the cytoplasm. This condition is known as coenocytic. The coenocytic mycelium is a characteristic feature of *phycomycetes* (Fig. 3.1).

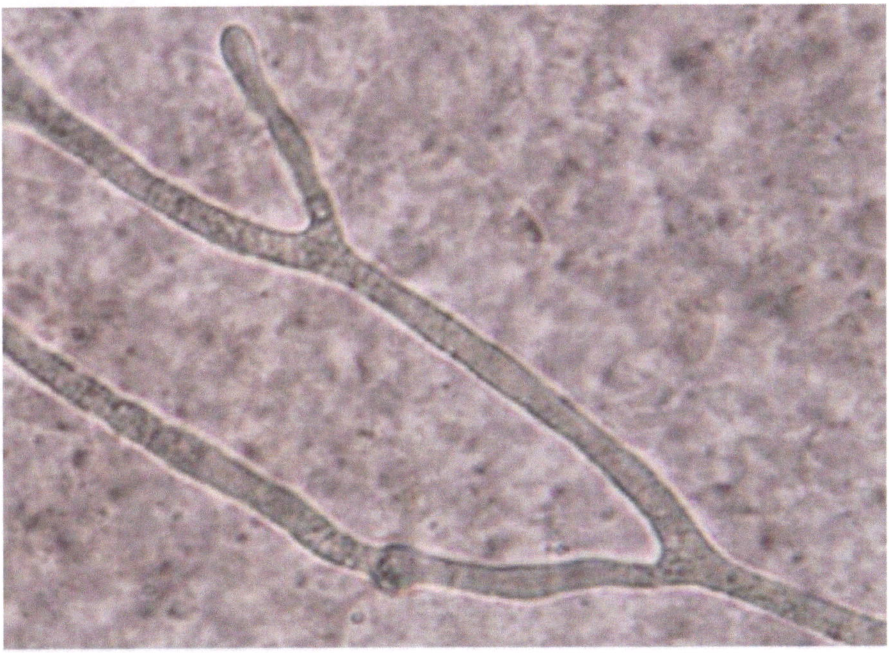

Fig. 3.1: Coenocytic or aseptate mycelium.

3.1.1.2. Septate mycelium

Mycelium, in which the septa, or cross walls, divide the hyphae into numerous uni- or multi-nucleated cells. A septum is formed by an annulus of wall material, which invaginates the plasma membrane and grows centripetally inward. Septate mycelium is a characteristic feature of higher fungi (*Ascomycetes* and *Basidiomycetes*). *Deuteromycetes* fungi also have septate mycelium (Fig. 3.2).

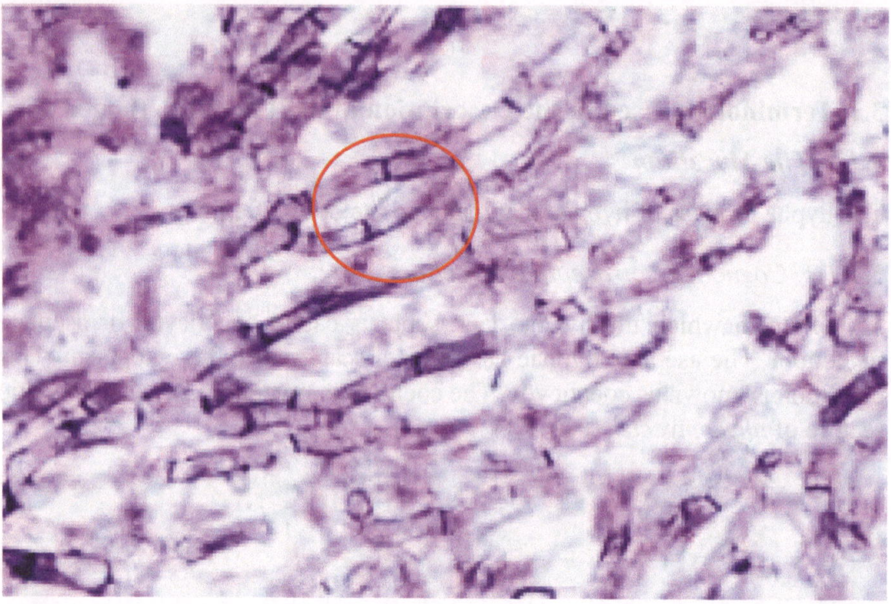

Fig. 3.2: Mycelium with septum.

3.1.1.3. Budding structures in yeast fungi

Budding is a common method of reproduction in *Saccharomyces* spp. The unicellular thalli put forth one or two buds that enlarge gradually. The protoplasmic contents, including mitotically divided nuclei, pass into the bud, and then the buds are cut off to form a new thallus. A small bud known as a bleb or daughter cell is formed on the parent cell. The nucleus of the parent cell splits into a daughter nucleus and migrates into the daughter cell. The bud increases in size while still attached to the parent cell and eventually breaks off and forms a new individual (Fig. 3.3).

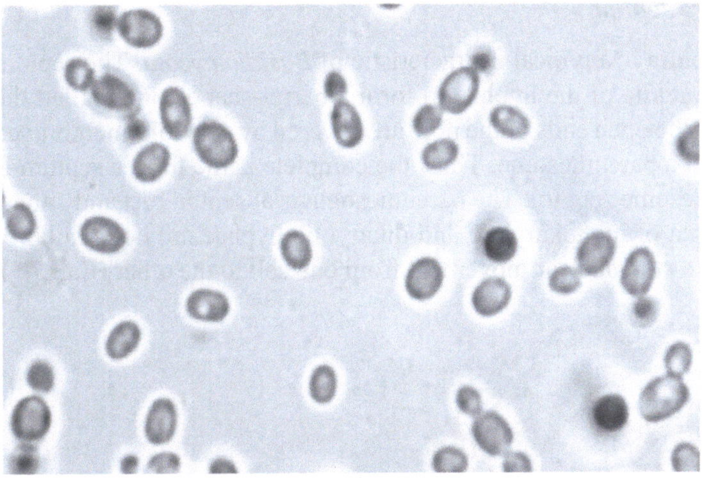

Fig. 3.3: Budding in yeast.

3.1.2. The Septum

In most fungi, hyphae are divided into cells by internal cross-walls called "septa" (singular septum). Two types of septums are observed in the fungi.

3.1.2.1. Perforated septum

These septa are usually perforated by pores large enough for transfer of nuclei, ribosomes, and mitochondria in between cells. It is present in most of the fungal division except *Basidiomycota* (Fig. 3.4).

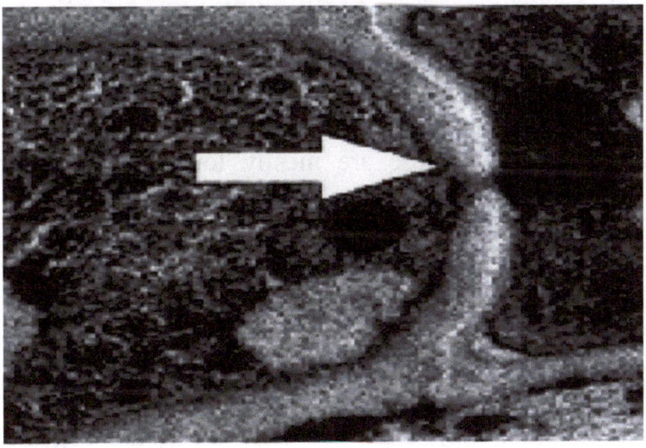

Fig. 3.4: Simple pore or perforated septum.

3.1.2.2. Dolipore septum

The dolipore septum is a typical characteristic of *Basidiomycota*. The septum flares up in the middle of the hyphae to form a barrel-shaped pore called the "dolipore." Both the open ends of the pore are covered by a double-membraned "pore cap" called a parenthesome. Thus, the complete name for the septum is dolipore-parenthesome septum. The parenthesome is absent in rust and smuts. The dolipore septa of monokaryotic and dikaryotic hyphae are not similar. It is a mechanism to stop nuclear movement from one cell to the other (Fig. 3.5).

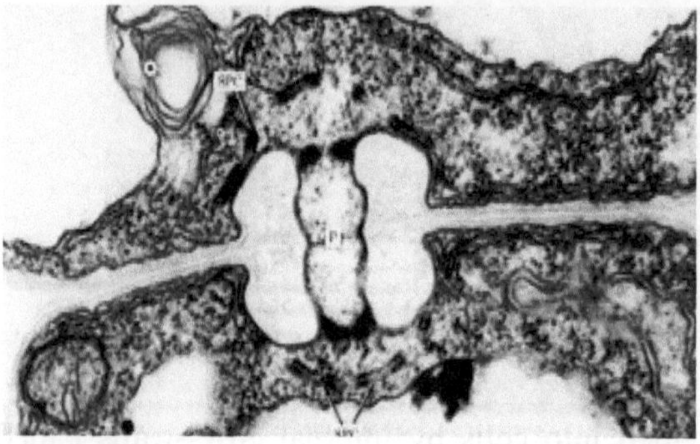

Fig. 3.5: Dolipore septum.

3.1.3. The Fungal Sex Organs

3.1.3.1. Sex organs

The sexual organs of fungi are called gametangia. Individual sex cells are called gametes and may contain one or more nuclei. The morphologically similar gametes are called isogametes, which are indistinguishable. The morphologically dissimilar gametes are called heterogametes, which can be differentiated into male and female gametes based on their differentiated structures. In this case, male gametangium is called antheridium and female gametangium is called oogonium. These are mostly found in the fungi *oomycetes*.

a. Antheridium

Antheridium is an elongated, club-shaped, multinucleate cell that is much smaller in size than oogonium and is produced terminally or intercalary.

These are differentiated based on their formation on the mycelium thallus.

(a) Monoclinous: Antheridium may arise anywhere along the oogonial stalk. e.g., *Pythium* sp.

(b) Diclinous: Antheridium arises from another branch of the thallus. e.g., *Pythium* sp.

(c) Paragynous: The tip of the antheridium is attached parallel to the side of the oogonium during gametangial contact. e. g. *Pythium* sp., *Albugo* sp.

(d) Amphigynous: Oogonium perforates the walls of antheridium and penetrates it finally emerging out at the other side as a special structure. On maturity, the sexual apparatus consists of a balloon-shaped oogonium with a thick-walled oospore at the top and a funnel-shaped base formed by the antheridium. e.g., *Phytophthora* sp.

b. Oogonium

The female sex organ, oogonium, is a globose structure that develops from the hyphae's terminal or intercalary swellings. The uninucleate dense central region of oogonium is called an oosphere or ooplasm and an outer or peripheral multinucleate spongy portion lining the oogonial wall is called as periplasm (Fig. 3.6).

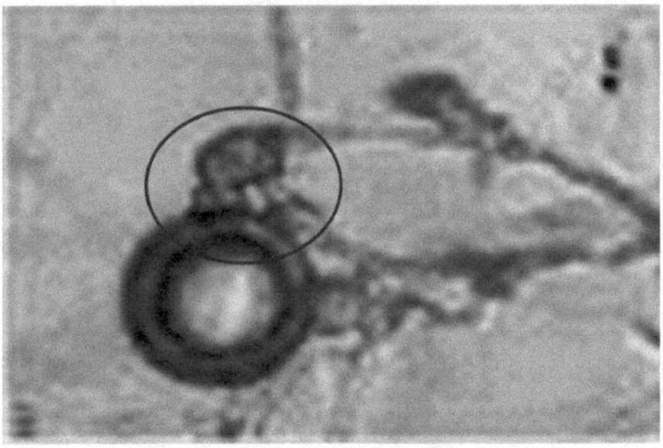

Fig. 3.6: Antheridium and Oogonium as a fungal sex organ.

On the basis of the distribution of sexual organs, fungi are classified as given below:

1. Hermaphrodite

   In this group of fungi, both male and female sex organs occur on the same thallus.

2. Dioecious

   Both male and female sex organs are present on different thalli.

3. Sexually undifferentiated

   The male and female sex organs are morphologically similar and therefore indistinguishable.

3.1.3.2. Mating types

The sexual reproduction in fungi occurs by the following methods:

Plasmogamy

In plasmogamy, anastomosis of two cells or gametes occurs and fusion of their protoplasts takes place. In this process, the two haploid nuclei of opposite sexes in the gametes are brought together, but these nuclei do not fuse.

Karyogamy

The fusion of two haploid nuclei brought together as a result of plasmogamy is called karyogamy. This process follows immediately after plasmogamy in many of the lower fungi or may be delayed in higher fungi.

In higher fungi, plasmogamy results in a binucleate cell containing one nucleus from each gamete or cell. Such a pair of nuclei is called a dikaryon. These two nuclei may not fuse. Meanwhile, during growth and cell division of the binucleate cell, the dikaryotic condition may be perpetuated from cell to cell by conjugate division of the two closely associated nuclei and by the separation of the resulting sister nuclei with two daughter cells. Nuclear fusion, which eventually takes place in all sexually reproducing fungi, is followed by meiosis.

Meiosis

Karyogamy results in the formation of a diploid (2n) nucleus. Meiosis (Gr. Meiosis = reduction) reduces the number of chromosomes to haploid and constitutes the third phase of sexual reproduction. This nucleus undergoes a reduction division to form two haploid nuclei each with "n" chromosomes. A mitotic division follows and four nuclei are formed. In *ascomycetes*, another nuclear division takes place resulting in the formation of eight nuclei. The nuclei get surrounded by a small amount of cytoplasm and secrete a wall to become spores. In a true sexual cycle, the above three phases occur in a regular sequence and usually at specified points. If there is only one free-living thallus, haploid or diploid in the life cycle of a fungus is called haplobiontic (Gr. haplos = single + bios = life) e.g. *Oomycetes* haploid gamete and diploid mycelium. If a haploid thallus alternates with a diploid, the life cycle is called diplobiontic (Gr. diplos = double + bios = life) e.g. *Allomyces* (water mold *Coelomomyces*, mosquito parasite) and in some yeasts.

Fungi achieve plasmogamy by a variety of methods:

i. Planogametic copulation

ii. Gametangial copulation

iii. Gametangial contact

iv. Spermatization

v. Somatogamy

i. Planogametic copulation

It is the fusion (or copulation) between gametes. Gametes are naked, wall-less sex cells that copulate to form a zygote. If two gametes are similar in size, they are called isogametes, and their copulation is called isogamy. Copulation between two dissimilar gametes, one smaller (male) and the other bigger (female), is called anisogamy. It occurs in the case of the genus *Allomyces*, a chytrid. The fusion between motile male gametes and non-motile female gametes (oosphere or egg) lying in the oogonium is called heterogamy. It occurs only in one genus, *Monoblepharidomycetes*. The motile gametes are called planogametes.

ii. Gametangial copulation

It is the fusion between gametangia (or the sex organs). When gametangia are similar in shape and size, these are called iso-gametangia and are designated as (+) and (−) gametangia rather than male and female. When the gametangia are different in shape and size, they are called hetero-gametangia. The male is usually smaller and club-shaped while the female is bigger and globular. Fusion between two similar gametangia results in a zygote which is called a zygospore. The zygote formed by the fusion between morphologically distinct gametangia is called oospore and the process oogamy. The plasmogamy between them is called gametangial copulation or contact.

Gametangial copulation is of two types:

The entire gametangia fuse, the intervening wall disappears and their contents come to lie in the common cell formed by their fusion. e.g. *Mucor*.

The contents of the male gametangia migrate into the female gametangium through a pore or fertilization tube and the male gametangium is left empty. e.g. *Rhizophydium*.

iii Gametangial contact

The male nuclei (and not the cytoplasm of antheridium) migrate into the oogonium through a pore dissolved at the point of contact or through a fertilization tube formed by the antheridium. e.g. *Pythium*, *Phytophthora*, *Albugo*.

iv Spermatization

It occurs in *Ascomycota* and *Basidiomycota*. Some fungi, like *Puccinia graminis tritici*, bear numerous minute, non-motile, uninucleate, male cells called spermatia (sing. spermatium; Gr. spermation=little seed). Spermatia are produced in spiral receptacles called spermagonia (sing. spermagonium) or pycnia (sing. pycnium). Insects, wind, or water flashes carry them from the female gametangium to a special receptive hypha (or trichogyne) to which they get attached. A pore develops at the point of their contact, and the spermatium releases the male nucleus into the female gametangium through a pore. This results in plasmogamy and the initiation of the dikaryotic stage of the cell. Spermatia are formed like conidia on spermatiophores. The spermatiophores may be formed exogenously or inside a spermogonium, e.g. *Puccinia*.

v. Somatogamy

Somatogamy (Gr. soma = body + gamos = marriage, union). In this, sex organs are not formed and somatic cells as such act as gametangia and fuse together e.g. *Agaricus*. Somatogamy may occur between cells of the same hypha (in a homothallic fungus) or between cells of different thalli (in a heterothallic fungus). Anastomosis, which is the fusion of hyphae is frequent in *Ascomycota* and *Basidiomycota* (Fig. 3.7).

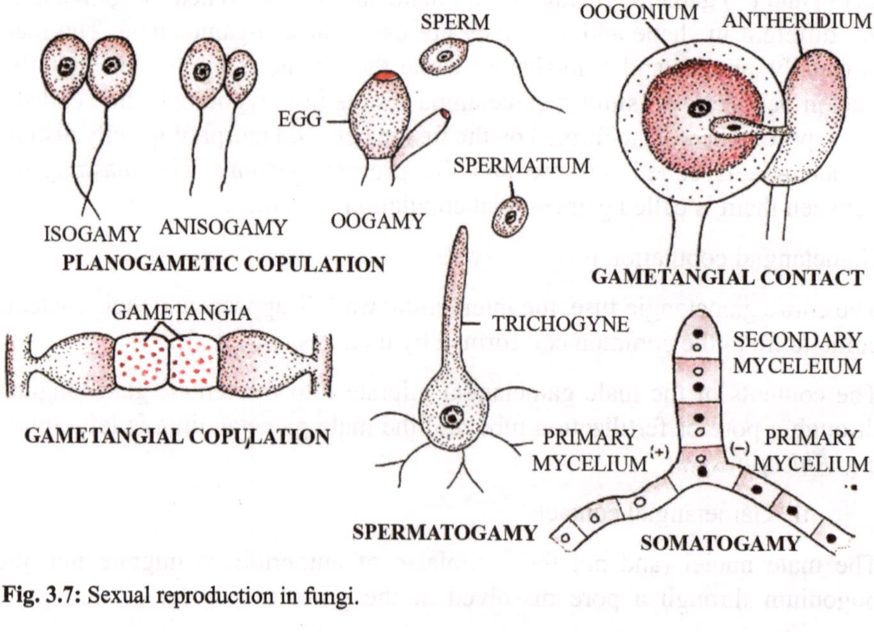

Fig. 3.7: Sexual reproduction in fungi.

3.1.4. The Fungal Spores

The small propagative or reproductive unit of the fungus (which separate from a hypha or a sporogenous cell) which gives rise to a new individual identical to the parent fungal species, is called a spore. This spore germinates with the emergence of a germ tube from them whose growth develops into hyphae and mycelium. The characteristics of spores are the absence of cytoplasmic movement, less water content, absence of vacuoles, and slow metabolism.

These fungal spores may be asexual fungal spores or sexual fungal spores depending on whether they are reproduced vegetatively or sexually.

3.1.4.1. Asexual fungal spores

These are also known as mitospores. Fungal spores are produced asexually, either in a specialized sac-like structure or on the tip or side of a fungal sporophore or conidiophore. The sac-like structures where the asexual spores are produced are known as sporangia, and the tiny spores within these sporangia are called sporangiospores. These sporangiospores may be either motile or non-motile, depending on the fungal genera where they are formed (Fig. 3.8).

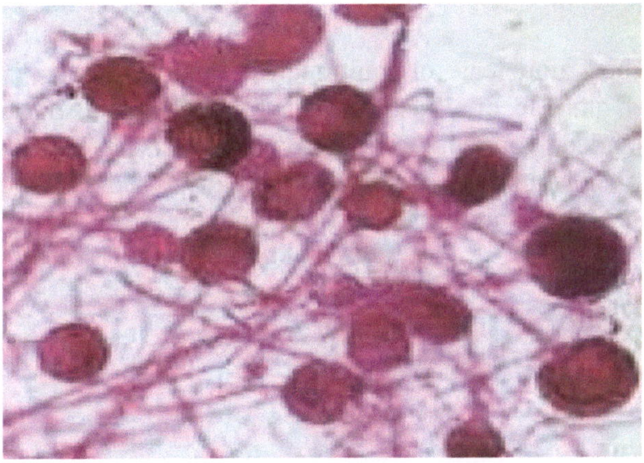

Fig. 3.8: The sac-like structure of sporangium.

Based on the motility of the sporangiospores (motile and non-motile) these are named as:

a. Planospores

These are motile spores known as zoospores. These zoospores are equipped with one or two flagella *viz.* whiplash and tinsel types of flagella. These zoospores are commonly found in the lower fungi.

b. Aplanospores

If sporangiospores are non-motile without flagella, they are called aplanospores. They are either uni-nucleate or multinucleate, formed in the sporangium surrounding the empty center or base called the columella, e.g. *Mucor* sp., *Rhizopus* sp. They are commonly found in the phylum *Zygomycota*.

The vegetatively produced spores at the tips or sides of hyphae in various ways are called conidia.

The different asexual spores in the fungi are sporangiospores, conidia, and chlamydospores

a. Sporangiospores

Sporangiospores are formed in a sporangium and may be motile (planospores) or non-motile spores (aplanospores). In lower fungi that inhabit aquatic or moist terrestrial places, the sporangiospores are usually motile and are called zoospores. In the *Zygomycetes* and especially in the *Mucorales*, the non-motile asexual spores called aplanospores are contained in globose sporangia surrounding a central core, or columella. Sporangia without the presence of columella is also known, where the spores (aplanospores) are arranged in a row inside a cylindrical sac termed a merosporangium (e.g. *Syncephalastrum* spp., *Mucorales*).

These aplanospores may be uninucleate or multinucleate and are generally unicellular, smooth-walled, globose, or ellipsoid in shape. When aplanospores mature, they may be surrounded by mucilage and rain splash or insects usually disperse such spores. When aplanospores are dry, they are dispersed by wind currents. The sporangiospores in sporangium may vary in number from several thousand to only one (Fig. 3.9).

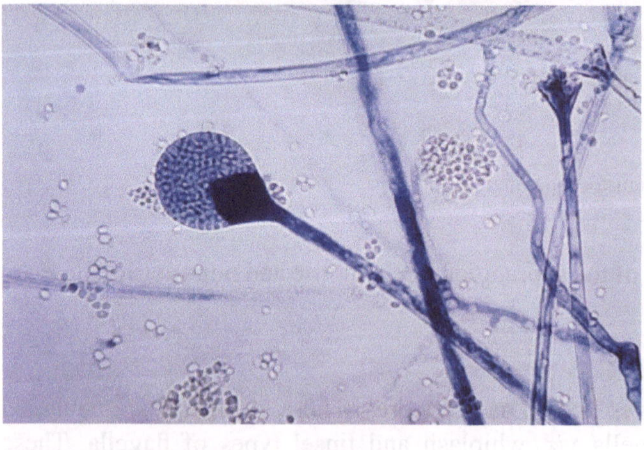

Fig. 3.9: Sporangiospores in sporangium.

In some fungi, a few spores containing sporangia are called sporangiola. Sporangiola are dispersed as a unit, e.g. *Choanephora* sp. and *Blakeslee* sp. in *Choanephoraceae* of *Mucorales*. In holocarpic thalli, the entire thallus (without differentiation of a sporophore) becomes a sporangium. Its contents cleave into a number of segments that round off and become zoospores. In eucarpic thalli, a part of the thallus, or special branches from the thallus, function as or produce sporangia.

In terrestrial and plant parasitic forms of lower fungi, the sporangium may function as a spore, and no zoospores are formed. In others, zoospores are formed within the sporangium itself or the inner wall of the sporangium may grow out into a short or long tube that swells to form a vesicle. The contents of the sporangium move into a vesicle where it is differentiated into zoospores e.g. *Pythium aphanidermatum.*

b. Zoospore

It is an asexually produced spore that is motile by means of a flagellum or flagella. Zoospore is naked, and its covering is only a hyaloplasm membrane. Normally, zoospores are uninucleate and haploid. Zoospores may be spherical, oval, pyriform, obpyriform, elongate, or reniform in shape. The zoospores are provided with one or two flagella (sing. flagellum) for its movement in the surrounding film of water.

Flagellum is a hair-or tinsel like structure that serves to propel a motile cell. These flagella may be anterior, posterior, or laterally attached to a groove in the body. There are two types of flagella in zoospores. They are whiplash and tinsel types. The whiplash flagellum has a long rigid base composed of all the eleven fibrils and a short flexible end formed of the two central fibrils only. The tinsel flagellum has a rachis, which is covered on all sides along its centre length with short fibrils. In uniflagellate zoospores, the flagellum may be anterior or posterior. But in biflagellate zoospores one is whiplash and the other is tinsel type and one points forward and the other backward. But in *Plasmodiophorales* fungi flagella are of whiplash type and unequal (Fig. 3.10).

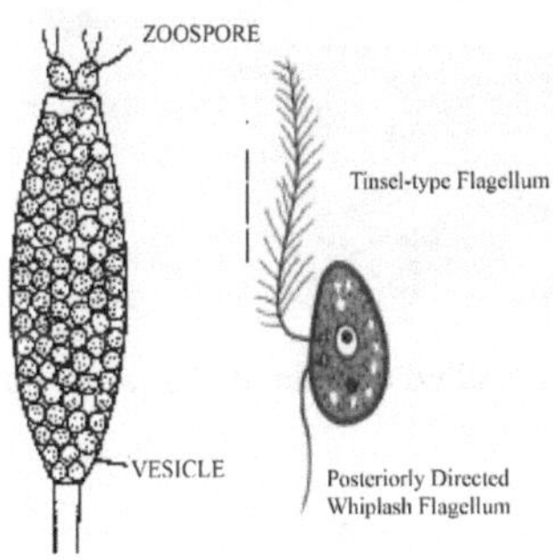

Fig. 3.10: Motile zoospore with different flagellation.

Zoospores have three phases in their life cycle, *viz.*, motility, encasement, and germination. The distance of their motility depends on available moisture, temperature, and the presence of stimulatory or inhibitory substances in the environment. Later on, the zoospores become sluggish to cast their flagella (except in *chytridiacious* fungi and primary zoospores in *Saprolegniales*), where flagella are shed and the zoospore body becomes spherical, secretes a thin wall around itself, and become encysted. The encysted zoospores germinate. The functions of zoospores include the initiation of new generations and acting as gametes.

c. Conidiospores or true conidia

Conidiospores or conidia (sing. conidium), are asexual reproductive structures borne on special spore-bearing hyphae known as conidiophores. They are found in many different groups of fungi, but especially in *Ascomycotina*, *Basidiomycotina*, and *Deuteromycotina*.

In *Deuteromycotina*, conidia are the only means of reproduction. Conidia may be borne singly, or in chains, or in clusters. They vary from unicellular (e.g. *Colletotrichum*), bicellular, microconidia of *Fusarium* spp., and multicellular (*Pestalotiopsis, Cercospora*).

One-celled spores are called amerospores, two-celled spores are called didymospores, and multicellular spores are called phragmospores.

The multicellular conidia may be divided by the septa in one to three planes. In *Alternaria* spp., conidia are with both transverse and longitudinal septa are called dictyospores. In species of *Fusarium,* one or two-celled microconidia and many-celled macroconidia are common (Fig. 3.11).

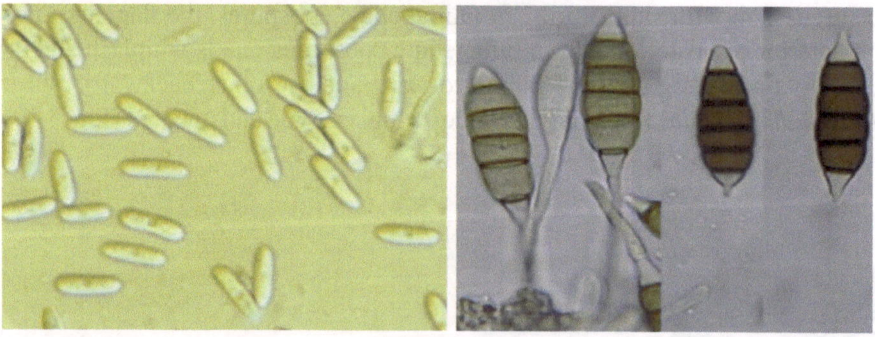

Fig. 3.11: Single-celled Conidia of *Colletotrichum* sp. and multicelled conidia of *Pestalotia* sp.

The shape of the conidium may vary. They may be globose, elliptical, ovoid, cylindrical, branched, or spirally coiled, or star-shaped (staurospores). The color of the conidia may be hyaline (hyalospore) or colored (phaeospore) pink, green, or darkly pigmented. The dark pigments are probably due to its melanin content.

The colour of the conidia and conidiophores are important features used in classification of the fungal genus in some orders.

Conidia may be formed in acropetal (oldest conidium at the base and the youngest at the apex) or basipetal (oldest conidium at the apex and youngest at the base) succession. Generally, the term "conidia" is used for any asexual spores other than sporangia and spores formed directly by hyphal cells.

When the spore is not much differentiated from the cells of the conidiophore in shape, the term "oidium" is often used for such conidia (Fig. 3.12).

Fig. 3.12: Oidium similar to conidiophore cells.

A distinction between sporangiospores and conidia is that, before germination of sporangiospores, a new wall, eventually continuous with the germ tube, is laid down within the original spore wall, whereas in conidia there is no new wall layer laid down. Conidiophores are also known as sporophores. They are special hyphae bearing conidia.

Conidiophore may be free, simple, or branched. They may be distinct from each other or may be aggregated to form compound sporophores or fruiting bodies such as synnemata, sporodochia, acervuli, and pycnidia. They may be provided with sterigmata or specialized branches on which they bear conidia. Some conidial spores are inflated at the tips (e.g. *Aspergillus*); others are inflated at intervals, forming knee-like structures on which the conidia are grouped (*Gonatobotrys*); still others have many branches, which are characteristically arranged, in whorls (*Verticillium*) or in sympodiums (*Monopodium*).

They are generally produced on the surface of the host. The sporogenous part of the conidiophore is commonly apical but may be laterally placed. The apical zone of differentiation of conidiophore may give rise to a single conidium or, more often, to a succession of conidia in chains, or false heads (Fig. 3.13).

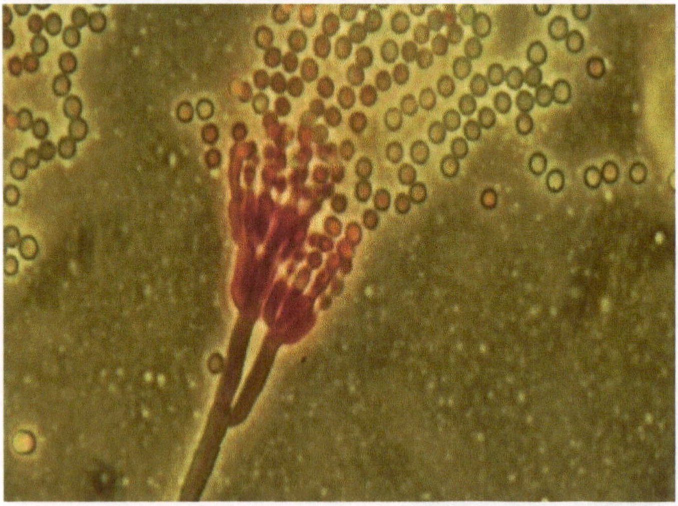

Fig. 3.13: Special structure (Phialides) bearing conidia in *Penicillium* sp.

c. Chlamydospores

Chlamydospore is a thick-walled spore formed due to the conversion of mycelial cells into the spore due to unfavorable climatic conditions. It generally functions as a resting spore. Chains of round and thick-walled spores are formed on the hyphae. When the rest of the hyphae die and disintegrate, the chlamydospores survive and later germinate to give rise to the fungus, e.g. in *Fusarium*, *Mucor racemosus*, *Saprolegnia* sp.

When chlamydospores are found in between fungal cells they are called "intercalary chlamydospores." Chlamydospores produced at the apex of the hypha are called "apical or terminal chlamydospores." During the formation of chlamydospores, the terminal or intercalary segments of the mycelium may become packed with food reserves and develop thick walls. The walls may be colorless or pigmented with dark melanin pigment.

Generally, there is no mechanism for detachment and dispersal of chlamydospores. They become separated from each other by the disintegration of intervening hyphae. They are the important reproductive organs for asexual survival in soil fungi (Fig. 3.14).

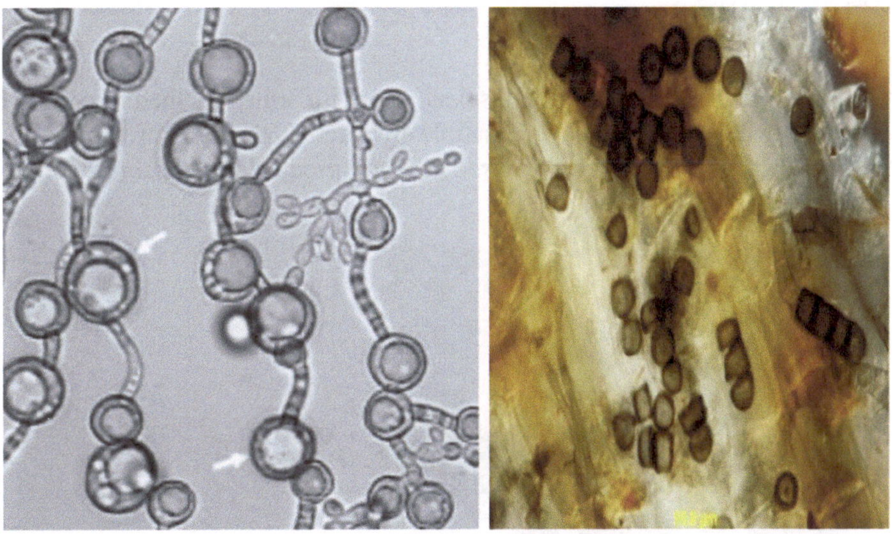

Fig. 3.14: Conversion of mycelial cells into chlamydospores and chlamydospores in root tissue.

3.1.4.2. Sexual fungal spores

The sexual spore of fungi is of two kinds:

1. Diploid resting spore e.g. oospore, zygospore.
2. Haploid non-resting spore e.g. ascospore, basidiospore.

1. Oospore

They are the thick-walled resting diploid sexual spores produced by the members of *oomycetes*, through gametangial contact between the male gamete antheridium and female gamete oogonium. These may be hyaline to golden yellow. These are produced in the downy mildew disease causing fungi (Fig. 3.15).

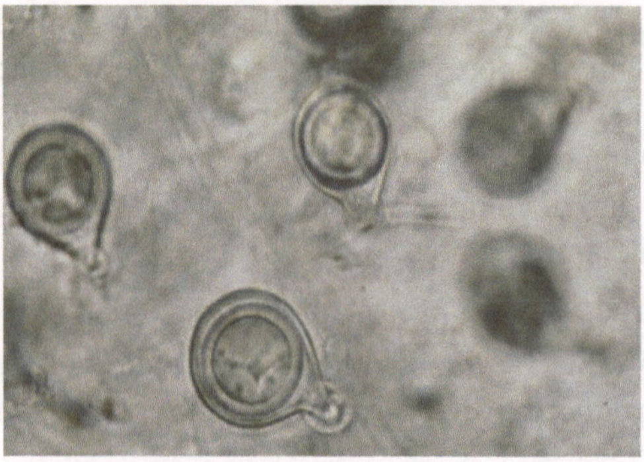

Fig. 3.15: Oospores.

2. Zygospore

The thick-walled resting, diploid, sexual spore is formed by the union of two similar sexual cells of two compatible hyphae. Zygospores are formed in the fungi of the class *zygomycetes*. These are produced in *Rhizopus* soft rot diseases (Fig. 3.16).

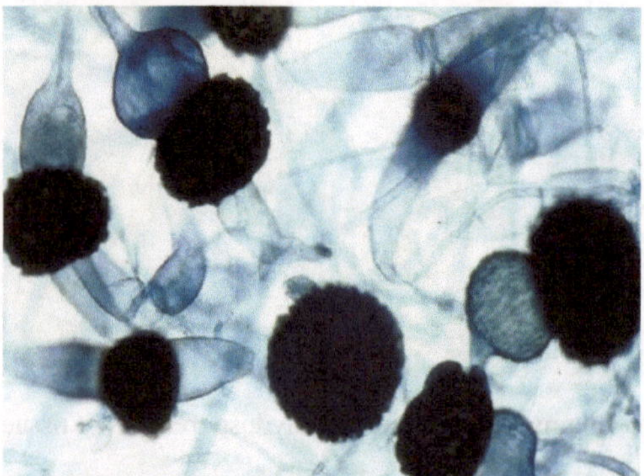

Fig. 3.16: Zygospores.

3. Ascospore

These are sexual spores produced in the fungal class *Ascomycota.* These spores are generally non-motile and produced in a sac-like structure known as an ascus. These are generally observed in powdery mildew diseases, where the ascus is formed in a closed fruiting body known as cleistothecia. The number of ascospores in the ascus may vary depending on the fungal genus. In some diseases, like *Taphrina* leaf spot or curl the ascus is produced as the naked layer of asci without fruiting bodies (Fig. 3.17).

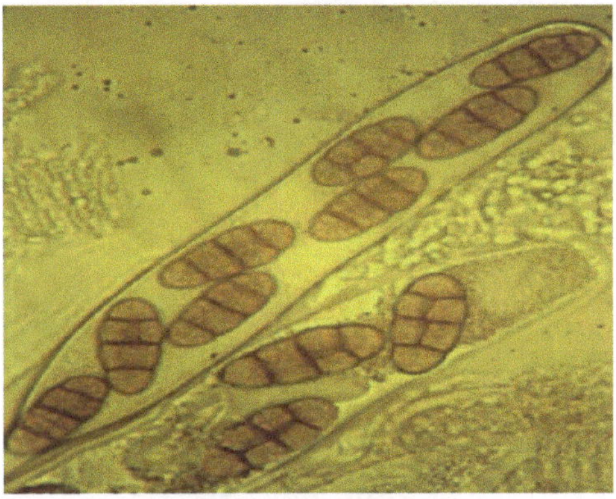

Fig. 3.17: Asci carrying ascospores.

4. Basidiospore

A sexual spore produced by *basidiomycotina* is known as a basidiospore. Basidiospores are produced at the tip of four sterigmata present on a club-shaped structure called basidium. These are typically produced in rust fungi (Fig. 3.18).

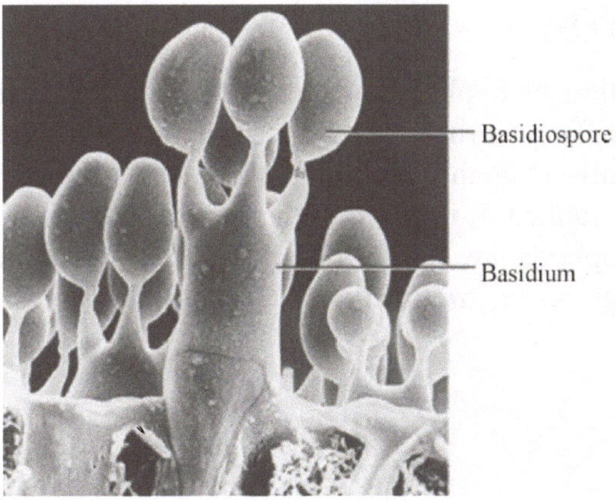

Fig. 3.18: Basidiospores.

3.1.5. The Fungal Fruiting Bodies

3.1.5.1. Asexual fruiting bodies

a. Synnema (Pl. synnemata)

Synnema are clusters of conidiophores and conidia that come up from infected tissue. It is the loose aggregation of the erect conidiophore that forms a dense fascicle, similar to the mycelial strand (Fig. 3.19).

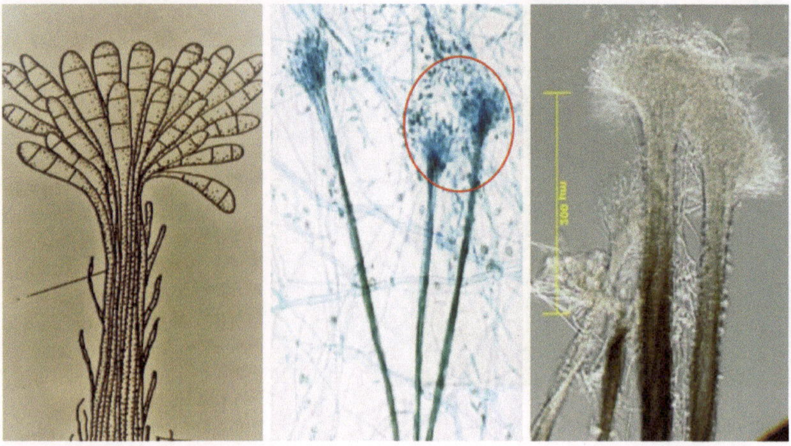

Fig. 3.19: Synnema (microscopic view).

The dense fascicle may split in different ways near the apex, sometimes resembling a feather duster. Such an arrangement is called coremium, and the conidia are produced at its apex, e.g. *Ceratocystis* sp., *Graphium* sp.

b. Sporodochium (Pl. Sporodochia)

It is a cushion-shaped aggregation of hyphae that breaks through the host surface and bears conidiophores. These structures may also be formed in the mass of hyphae lying superficially over the substrate. It is a fruiting body peculiar to acervuli, but it is not sunken. It is raised in the form of a cushion-shaped stroma on which conidiophores are borne. These can be observed in *Monilinia fruiticola* and *Claviceps purpurea* infections (Fig. 3.20).

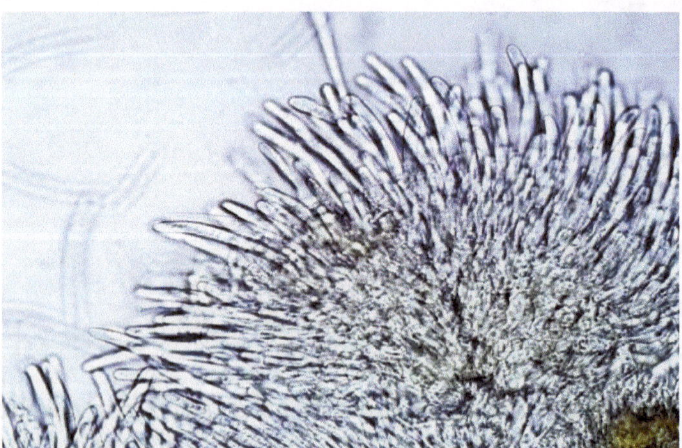

Fig. 3.20: Conidiophores produced from sporodochium on apple bark.

c. Acervulus (Pl. Acervuli)

Most of the fungi belonging to the order *Melonconiales* of the *Deuteromycotina* produces acervuli. It is a saucer-shaped, depressed mass of aggregated hyphae bearing conidiophore in a compact layer on its exposed surface. In between the conidiophore, long, pointed, dark-coloured structures called setae are present. On the host, acervuli are initially developed below the cuticle or epidermis and become erumpent on maturity, e.g. *Colletotrichum* sp., *Pestalotia* sp. These are mostly observed in anthracnose diseases (Fig. 3.21).

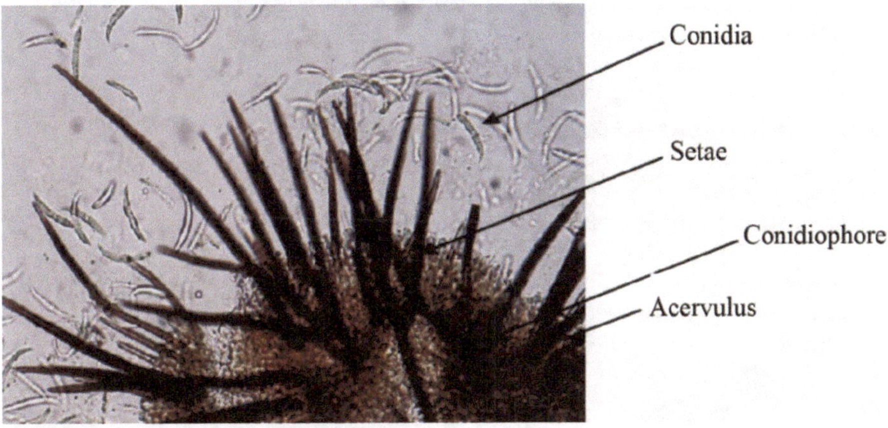

Fig. 3.21: Acervulus having setae and conidia.

d. Pycnidium (Pl. pycnidia)

It is a hollow, flask, or globose-shaped fruiting body with an opening known as the ostiole, in which pseudo-parenchymatous inner walls (peridium) are lined with conidiophore-bearing conidia, e.g. in *Macrophomina phaseolina*, *Botrydiplodia theobromae*, *Diplodia natalensis*. These are observed in leaf spot diseases (Fig. 3.22).

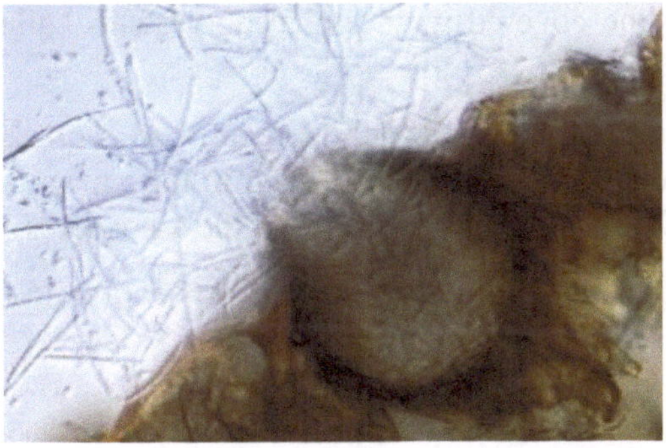

Fig. 3.22: Pycnidium with ostiole and release of pycniospores.

e. Sorus (Pl. Sori)

The spore-bearing hyphae are grouped into small to large masses or clusters of spores covered with a protective thin silvery or colored membrane, e.g. smut or rust sori (Fig. 3.23).

Fig. 3.23: Seeds entirely replaced by fungal sori.

3.1.5.2. Sexual fruiting bodies

a. Cleistothecium

A cleistothecium is a globose or round, completely closed fruiting body of the fungus with no special opening to the outside. The ascomatal wall is called the peridium and typically consists of densely interwoven hyphae or pseudoparenchyma cells. It may be covered with hyphal outgrowths called appendages.

The asci inside the cleistothesium are globose, deliquescent, and scattered throughout the interior cavity i.e. as in *Eurotium* or arising in tufts from the basal region of ascocarps as in *Erysiphe*. These asci contain the ascospores.

Cleistothecia are the sexual fruiting bodies mostly formed in powdery mildew fungi (Fig. 3.24).

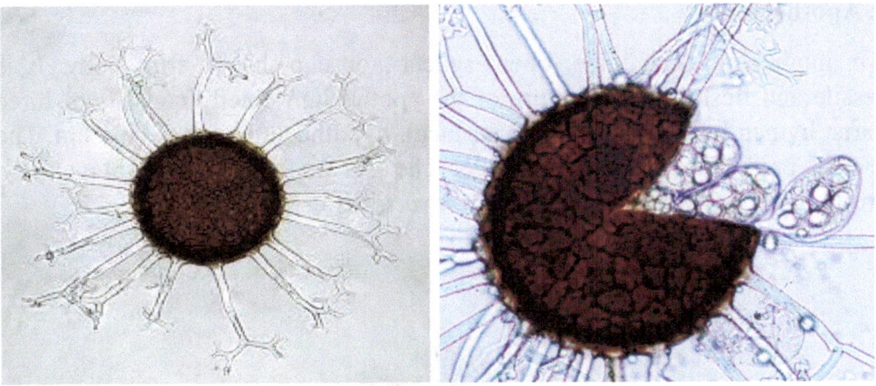

Fig. 3.24: Cleistothecium of powdery mildew pathogen *Phyllactinia* sp. with appendages.

b. Perithecium

These sexual fruiting bodies have flask-shaped structures with an opening by a pore or ostiole (short papilla opening by a circular pore) through which the ascospores escape. The ostiolar canal may be lined by hair-like structures called periphyses. The unitunicate asci are usually cylindrical, borne on a stipe (stalk), and released from a pore.

The perithecial wall is composed of fungal tissue called pseudoparenchyma, which are observed in fungal order Sphaeriales and Hypocreales. Perithecia are also found in the genus Xylaria (Dead Man's Fingers, Candle Snuff), *Venturia,* Nectria, Claviceps, and Neurospora (Fig. 3.25).

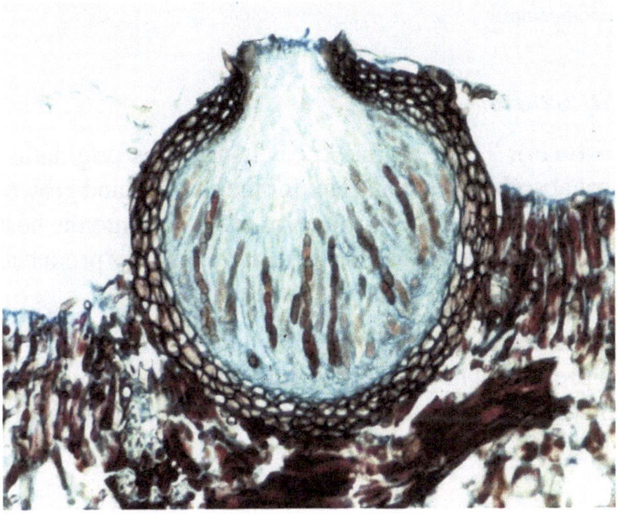

Fig. 3.25: Perithecium of *Venturia* sp. causing apple scab.

c. Apothecium

An apothecium is a wide, open, saucer-, or cup-shaped fruit body. It is sessile and fleshy. The structure of the apothecium chiefly consists of three parts: hymenium (upper concave surface), hypothecium, and excipulum. The asci are present in the hymenium layer. The asci are freely exposed at maturity. An example is the members of *Dictyomycetes* (Fig. 3.26).

Fig. 3.26: Vertical section of apothecium.

3.1.6. Special Fungal Structures

a. Haustorium (pl. Haustoria): These are the special structures originating from the fungal hyphae that absorb the host nutrient for the survival and growth of the fungus. It can be a simple or branched projection of hyphae into the host cells, may be knob-shaped or branched finger-shaped, and is mostly produced as intracellular absorbing structures of obligate parasites.

Haustoria are usually produced in those fungi in which the formation of intercellular mycelium is present. They secrete certain specific enzymes that hydrolyze the proteins and carbohydrates of the host cell, and thus they absorb nutrients from the host (Fig. 3.27).

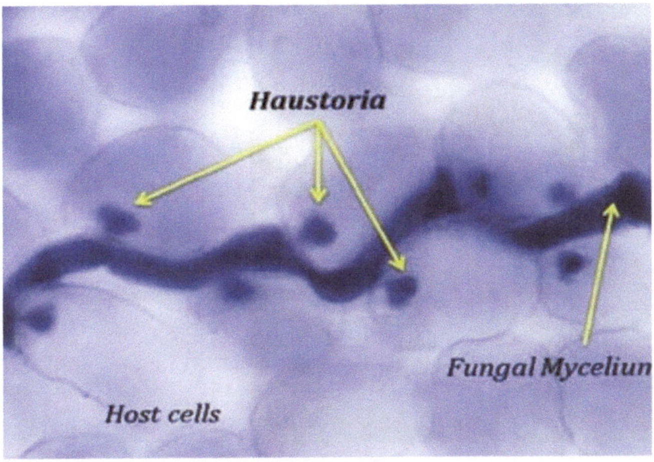

Fig. 3.27: Haustoria formed within host cell.

b. Appressorium: An appressorium is a flattened, hyphal "pressing" organ, from which a minute infection peg grows that facilitates attachment and penetration of the host by a fungus (Fig. 3.28).

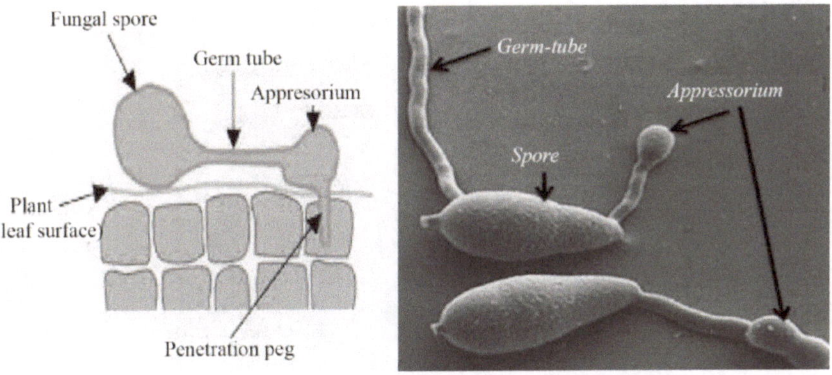

Fig. 3.28: Appressorium and penetration peg of *Oidium* sp. (causal organism of powdery mildew).

c. Sclerotia: The hyphae become interwoven to form a compact mass surrounded by a hard covering or rind and looks like the mustard seeds in the fungal growth. They remain dormant under unfavorable conditions and germinate into new mycelia on the return of favorable conditions, e.g. in *Rhizoctonia solani, Sclerotium rolfsii.* These are formed in root rot and collar rot diseases causing fungi. These can be observed on the infected plant surface or inside the host tissues (Fig. 3.29).

Fig. 3.29: Plant infected with *Sclerotinia sclerotiorum* having sclerotia on the infected portion.

d. Rhizomorphs: In some higher fungi, several hyphae may become interwoven to form rope-like structures called rhizomorphs. Under favorable conditions, they resume growth to give rise to new mycelia e.g. *Armillaria mellea* (Fig. 3.30).

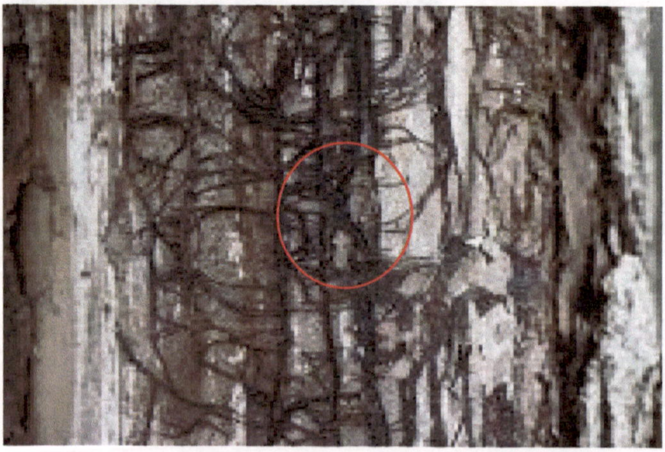

Fig. 3.30: Rhizomorphs of *Armillaria mellea* in infected plant.

e. Stroma: It is a compact somatic structure like a mattress, on which or in which fructifications are usually formed e.g. in Chestnut bight disease (Fig. 3.31).

Fig. 3.31: Stroma formed in chestnut blight disease.

3.2. Terminology Used in the Description of Fungal Disease, Infection, and Symptoms

3.2.1. Damping Off

The fungal pathogen may attack the pre-emerging or post-emergence tender and succulent seedling at any stage in the nursery or field. Sprouting seedlings get infected and wither before emergence from the soil (pre-emergence damping off). Water-soaked minute lesions appear on the stem near the soil surface, soon girdling the stem and spreading up and down in the stems. Within one to two days stem may rot, leading to the toppling over of the seedlings (post-emergence damping off). The young seedlings in the nursery are killed in patches, and infection spreads quickly. Under favourable conditions, the entire seedlings in the nursery are killed within 3 to 4 days. A thick weft of mycelium may be seen on the dampened seedlings near the soil surface.

The pathogens responsible for this disease are species of *Pythium, Phytophthora, Botrytis, Fusarium, Cylindrocladium, Diplodia, Phoma, Alternaria*, etc. besides *Rhizoctonia solani* and *Aphanomyces cochlioides* (Fig. 3.32).

Fig. 3.32: Symptoms of damping off seedlings.

3.2.2. Gummosis

Oozing of gum-like substances, due to fungal infection, from the main trunk of the plant just above ground level is known as gummosis. The disease is observed on the main trunk of the susceptible tree species, particularly sweet oranges and mandarins, just near or above ground level.

The first symptoms are the dark staining of infected and wounded trunk bark, which progresses into the wood. The infection starts at growth cracks or ridges at crotches. The bark is killed, and gum oozes out and dries, turning black. Exudation may stop after some time. They may be confined to certain seasons of the year. Large limbs are killed, and if left unchecked, the whole tree may be killed in the course of time. Sometimes branches break at the infected portion. The pathogen responsible for this disease is *Phytophthora* sp. (Fig. 3.33).

Fig. 3.33: Symptoms of gummosis disease in citrus tree.

3.2.3. Root Rot

The infection of the root leads to its rotting. The disease occurs at the seedling stage or other growth stages of the plant. When the pathogen infects the seedling, a dark brown lesion develops on the stem near ground level. The infected seedling slowly turns yellow, and the petioles and leaflets show drooping, leading to the complete drying of the plant. The stem near the collar region and below it exhibits rotting symptoms. The black dead tissues contain pinkish mycelial growth and sclerotia.

When the plant is infected during its growth stages, the infected plants appear in patches in the field. Leaves show yellowing, bronzing, and a slight wilting. Later, they turn brown and dry but remain attached to the plant. Below the soil line, and in some plants above the soil line, the bark and cambium turn brown, resulting in a firm brown rot of the root and the lower stem. The rotted roots are usually partly covered by coarse, brown, parallel strands of mycelium. When the plant is pulled out, the roots appear decaying and dark in color e.g. root rot of pulses caused by *Macrophomina phaseolina* (Fig. 3.34).

Fig. 3.34: Symptoms of dry root rot on chickpea.

3.2.4. Wilt

The drying and death of the whole plant due to the pathogen's infection is known as wilt. Plants infected at early stages remain severely stunted. The first symptoms can be seen as the bronzing of leaf veins, followed by interveinal chlorosis and the yellowing of leaves. Finally, the leaves begin to dry, giving them a scorched appearance. At this stage, the characteristic diagnostic feature is the drying of the leaf margins and areas between veins, which gives a "tiger stripe" or "tiger claw" appearance in cases of cotton wilt caused by *Verticillium* sp. The affected leaves fall off, leaving the branches barren. Infected stems and roots, when split open, show a vascular discoloration of the woody tissue, which may taper off into longitudinal streaks.

The disease is caused by the fungal genus *Fusarium, Verticillium, Ceratocystis, Ophiostoma, Leptographium* etc. (Fig. 3.35).

Fig. 3.35: Wilt symptoms on pomegranate plant due to *Ceratocystis*.

3.2.5. Collar Rot

Rotting at the collar region of the plant is known as collar rot. The pathogen can infect the plants at the seedling stage or at later stages of plant growth. When the seedlings are infected, it shows a circular, brownish spot on the hypocotyl region. The discolored area rapidly becomes soft and rotten and spreads onto the stems, and the seedling becomes yellow, rotten, and collapses. When the plant is infected at later stages, the fungus infects the plant at its collar region. The affected area becomes soft and rotten, with whitish mycelial growth on the collar region. The affected stems become shredded by the mycelium and sclerotial fruiting bodies of the pathogen.

The fungal pathogen responsible for the disease is *Sclerotium, Macrophomina, Aspergillus*, etc. (Fig. 3.36).

Fig. 3.36: Collar rot disease symptoms.

3.2.6. Wart

The appearance of wart-like protuberances on tubers and stems is known as a wart. The disease is most common in potato crops. A diagnostic symptom of wart disease is galls of irregular shapes produced on the stem and underground plant parts. There may be a reduction in vigor in the affected plants. Warts can be found on the stem near the ground level. Above-ground galls are green to brown, turning black at maturity and later decaying. The underground galls vary in shape but are mostly spherical, with corrugated surfaces, and range from pea-size to fist-size (<1 cm to >8 cm diameter). Below-ground galls are white to brown, turning black as they decay. These galls appear at stem bases, stolon tips, and tuber eyes. They may not be evident until harvest time. At harvest, galls may desiccate or decay. Tubers may be disfigured or completely replaced by galls. Tuber galls may also develop after harvest or in storage. The infected potato may not be killed, but the meristematic tissue of sprouts may be so severely attacked that plants may fail to emerge from such warted seed tubers. The disease is caused by the fungus *Synchytrium endobioticum* (Fig. 3.37).

Fig. 3.37: Wart-like protuberances on tubers and stem caused by *Synchytrium endobioticum.*

3.2.7. Club Root

The convergence of roots and rootlets of plants into small or large spindle-like spherical, knobby, or club-shaped swellings is termed club root. Plants infected by club root symptoms remain stunted, wilt easily, and may have yellowing of leaves. The roots of club-root-infected plants are swollen into thick, irregular club shapes. The disease is more common on the roots of cruciferous plants.

Club-root may develop extensively on plant roots before the first sign, i.e. an abnormal wilting and yellowing of leaves, especially on warm days, is noticed above ground. If the soil is moist, these symptoms may not become apparent until water stress occurs. When infection occurs at an early stage of growth, young plants are stunted and may die, whereas plants infected at a later stage fail to make marketable heads or grow. When diseased plants are pulled from the soil, the roots are usually swollen and distorted. Root malformation may vary in size from very small swellings on the tap and lateral roots to large club-shaped roots, depending on when the plants became infected. The fungal pathogen responsible for this disease belongs to the genus *Plasmodiophora* (Fig. 3.38).

Fig. 3.38: Club root of cabbage caused by *Plasmodiophora brassicae.*

3.2.8. Leaf Spot

Small, light brown, grey, or black-colored spots appear on the leaves, depending on the fungal pathogen responsible for the invasion. These spots increase in size, and coalesce to form large areas. These areas may turn necrotic or papery, depending on the invading fungal species. In some cases, the infected portions drop off, leaving a shot-hole symptom. Heavily infected leaves have premature defoliation. In severe cases, these spots also develop on petioles and stems. Most of the fungi in the *Deuteromycota* class cause leaf spot diseases (Fig. 3.39).

Fig. 3.39: Leaf spot caused by *Cercospora capsici on* capsicum leaves.

3.2.9. Leaf Blight

Leaf blights are generally longitudinal necrotic spots on the leaves. Under a humid environment, the size and diameter of the spots increase rapidly, giving a blighted appearance to the affected leaves and the crop. The blight fungus can attack seedlings as well as older plant leaves. Rapid browning and death of leaves give a burnt appearance to the plant, e.g. leaf blight of sorghum caused by *Exerohilum tursicum.*

The fungus responsible for leaf blight belongs to the genus *Helminthosporium, Cercospora, Phytophthora, Alternaria,* etc. (Fig. 3.40).

Fig. 3.40: Symptoms of leaf blight lesions on sorghum leaves.

3.2.10. Leaf Blast

The typical symptoms are spindle-shaped spots that can appear on leaves, leaf sheaths, rachis, nodes, and even the glumes. On the leaves, the lesions start as small, water-soaked bluish green flecks, soon enlarging, and forming characteristic spindle-shaped spots with a gray center and dark brown margins. The spots join together as the disease progresses, and large areas of the infected leaves dry up and wither.

Severely infected nurseries and fields show a burnt appearance e.g. blast of rice. The fungus genus responsible for leaf blast is *Pyricularia* (Fig. 3.41).

Fig. 3.41: Blast of rice.

3.2.11. Downy Mildew

The infection of downy mildew fungus appears as irregular, yellowish, translucent spots on the upper surface of the leaves, and the corresponding lower surface has milky white fungus growth. Affected leaves become yellow, which later turn brown and get dried. Premature defoliation occurs, followed by the dwarfing of tender shoots. The downy mildew fungus infects leaves, stems, inflorescence, and fruits. The infection on the stem causes brown, sunken lesions. On fruits and berries, the infection leads them to become leathery and shriveled. The fungal genera responsible for downy mildew symptoms are *Peronospora, Plasmopara, Sclerospora, Bremia, Pseudoperenospora, Phytophthora, Bremiella, Basidiophora, Paraperonospora, Sclerospora, Peronosclerospora, Sclerophthora*, etc. (Fig. 3.42).

Fig. 3.42: Downy mildew symptoms on grapevine leaf.

3.2.12. Powdery Mildew

The powdery mildew fungus attacks all the above-ground plant parts, *viz.* leaves, buds, stalks, inflorescence, and fruits. Powdery growth appears mostly on the upper surface of the leaves but can appear on the lower leaf surface as well. It manifests itself by the appearance of wefts of white mycelium on the affected parts. The whole surface is later covered with a powdery substance, which is blown away by even a slight disturbance caused by winds. Malformation and discoloration of affected leaves occur. The affected stem turns to dark brown discolouration. Floral infection results in the shedding of flowers and poor fruit set. Early fruit infection results in the shedding of affected fruits. The affected fruits do not grow in size and may drop before attaining pea size.

The fungus genera responsible for powdery mildew are *Erysiphae, Sphaerotheca, Leveillula, Uncinula, Phyllactinia, Blumeria, Podosphaera, Microsphaera*, etc. (Fig. 3.43).

Fig. 3.43: Powdery mildew symptoms on grape caused by *Erysiphe necator.*

3.2.13. Leaf Smut

The fungal pathogen produces slightly raised, angular, black spots (sori) on both sides of the leaves. Although rare, it can also produce smut sori on leaf sheaths. The black spots and sori are about 0.5 to 5.0 mm long and 0.5 to 1.5 mm wide. Many sori can be found on the same leaf, but these remain distinct from each other. Typical sori of leaf smut break open when wet and release black spores. Heavily infected leaves turn yellow, and the leaf tips die and turn gray. Leaf smut occurs late in the growing season and causes little loss. The fungus responsible for leaf smut belongs to the genus *Entyloma* (Fig. 3.44).

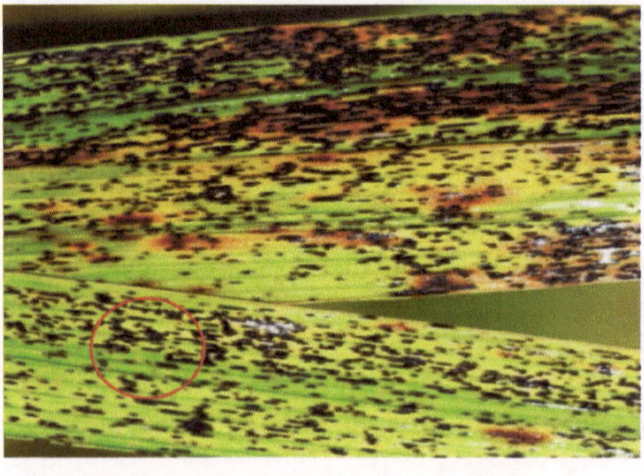

Fig. 3.44: Leaf smut of rice due to *Entyloma oryzae.*

3.2.14. Rust

The rust diseases are typified by the presence of rust sori or rusty-coloured pustules erupting through the infected plant surface, which may be leaf, stem, glumes, or grains. They can be easily distinguished from other leaf diseases by the eruption of rusty spore masses with a little scrap. Different colors of rust pustules, like brown, black, yellow, orange, or golden, can appear at the infection site of rust fungi, depending on the rust genus and species involved. The rust sori contain the rust spores, i.e. uredospore and teliospores. The rust diseases are caused by the fungus belonging to the order *Uredinales*.

The most common rust genera *are Puccinia, Cronartium, Gymnosporangium, Hemileia, Phakopsora, and Uromyces*, etc. (Fig. 3.45).

Fig. 3.45: Symptoms of rust on leaf and stem of a wheat plant.

3.2.15. Smut

Smut diseases are common in cereals and millet crops. The smut fungus infects the grains to convert them into smut sori that are filled up with the sooty black masses of the smut spores. Symptoms are usually noticed on the spikes in cereals and in grain earhead or cob in millet. Large smut sori replace the grains within, or sometimes the whole earhead. The smutted plants are stunted in growth and remain greener than the other plants in the case of maize, while in wheat, the infected plants are slender and the spike emerges earlier than the non-infected plants. Seeds in the infected ear head or entire ear head are replaced by sori, which are covered with a silvery-fine membrane containing a large number of black color, smut spores, or spore balls. e.g. loose smut of wheat caused by *Ustilago tritici.*

The most common genera causing smut disease are *Ustilago, Sporisorium, Tolyposporium*, etc. (Fig. 3.46).

Fig. 3.46: Symptoms of smut on wheat and maize.

3.2.16. Bunt

The bunt disease infects the grains in cereals, particularly wheat. The affected plant shows yellow leaf flecking at the late seedling stage, stunting, and excessive tillering prior to heading. Once the head emerges, the fungus transforms the kernels into spherical bunt balls. Infected heads may remain green for a longer period, and the spikelets and awns are at a wider angle on the main rachis. Bunt balls are filled with dark-colored spores that have a fishy odor. Bunt disease is caused by the fungus of the genus *Tilletia* (Fig. 3.47).

Fig. 3.47: Symptoms of bunt on wheat.

3.2.17. Malformation

There is a proliferation of infected tissues which may include floral malformation and vegetative malformation.

The flowering panicles instead of coming out as a normal one turns into just a compact bunch of hard flowers. An individual flower is greatly enlarged and has a large disc. The inflorescence gets hypertrophied. The percentage of bisexual flowers in malformed panicles is very low. There is no fruit setting in such a malformed flower head.

In vegetative malformation, the growing shoot form into a compact mass of leaflets which restricts the further growth of the shoot. The fungus responsible for malformation belong to the genus *Fusarium* (Fig. 3.48).

Fig. 3.48: Mango malformation.

3.2.18. Dieback

The stems start dying back from the tip downward. The symptoms include discoloration and darkening of the bark at some distance from the tip of the plant stems. Later, the dark area advances, and young twigs wither. This may be accompanied by the exudation of gum. The leaves of affected stems turn brown, and leaf margins roll upward. At this stage, the twig or branch shrivels, turns dark, is observed without leaves, is dead, and is ready to break or fall. The infected twigs show brown streaks of vascular tissues on splitting the twigs lengthwise.

The fungus responsible for this disease belongs to the genus *Colletotrichum, Botryosphaeria, Botrydiplodia, Nectria,* and *Phomopsis,* etc. (Fig. 3.49).

Fig. 3.49: Symptoms of dieback in mango plant by *Botrydiplodia themobromae*.

3.2.19. Anthracnose

The disease symptoms include leaf spots, wither tips, twig blight, blossom blight, and fruit rot, where the affected areas usually crack and sink slightly with the growth of fungal pathogens and fungal fruiting bodies, particularly acervuli. Infection on young leaves causes leaves to wither and dry up. Infection of leaf margins darkens the infected areas and dries up, giving the leaf a ragged appearance, and severely infected leaves may fall out. On tender twigs and branches, the infected area turns brown-black, and cracks with depression. Affected branches ultimately dry up. Similar symptoms appear on fruit also (Fig. 3.50).

Fig. 3.50: Symptoms of anthracnose on chili leaf and fruit.

3.2.20. Stem Rot

Stem rot disease is more common in rice cultivation. Initial symptoms are the appearance of small, irregular black lesions on the outer leaf sheath near or above the ground level. Lesions expand as the disease advances. Infected stems rot and numerous tiny white and black sclerotia and mycelium are found inside the infected culms. Infected culm lodges and develops unfilled panicles and chalky grain. Severe infection causes tiller death, and the disease aggravates the plants to lodge, e.g. stem rot of rice caused by *Sclerotium oryzae* (Fig. 3.51).

Fig. 3.51: Symptoms of stem rot on rice.

3.2.21. Bud Rot

The disease is common in palm trees. Palms of all ages are susceptible to this disease, but it is more severe in young palms of 5–20 years. The first indication of the disease is seen on the central shoot of the tree. The heart leaf shows discoloration, which becomes brown instead of yellowish brown. This is followed by drooping and breaking off the heart leaf. With the progress of the disease, more leaves lose their luster and turn pale yellow. The entire base of the crown may be rotten, emitting a foul smell. The central shoot comes off easily with slight pulling. The leaves fall in succession, starting from the top of the crown. The leaf-falling and bunch-shedding continue until a few outer leaves are left unaffected. But within a few months, the infection leads to the complete shedding of leaves, subsequent wilt, and death of the tree. The fungal pathogen of the genera *Phytophthora* is responsible for this disease (Fig. 3.52).

Fig. 3.52: Symptoms of bud rot on the coconut tree.

3.2.22. Leaf Sooty Mold

Sooty mold, also called blotch or black mold, appears as black coatings on leaves, stems, and fruit. The black residue of sooty mold is composed of dark fungal threads of a number of ascomycetes, besides species of Alternaria, Capnodium, Cladosporium, Fumago, and Scorias. These fungi grow in flowing sap or on honeydew excreted by aphids and other sucking insects. Black encrustation is seen on the upper and lower surface of the leaves. The superficial growth of the fungus affects photosynthetic activity (Fig. 3.53).

Fig. 3.53: Sooty mold on the upper leaf surface of sweet orange.

3.2.23. Leaf Galls

The gall develops as a chemically induced swelling arising from the surface of the leaf lamina, veins, midribs, and petiole. On the leaf lamina, it forms yellowish bulges standing out on the upper surface. On veins, midribs, and petioles, the gall appears as translucent yellow-white swellings that are often elongated and blister-like. It is generally caused by the fungus *Protomyces macrosporus* in coriander and by *Exobasidium camelliae* in camellias. The color of these galls in camellias can range from cream to light green to pink or reddish colour. As these galls mature, several layers of the lower leaf surface will peel away, revealing a white color, which are the spores of the fungus (Fig. 3.54).

Fig. 3.54: Symptoms of leaf galls on coriander leaf.

3.2.24. Stem Galls

The disease appears in the form of tumor-like swellings of leaf veins, leaf stalks, peduncles, stems, and fruits. The infected veins show a swollen, hanging appearance to the leaves. Initially, the tumors are glossy, but they rupture later on and become rough. They are about 3 mm broad and up to 12.5 mm long. Badly affected plants are killed. In the presence of excessive soil moisture, especially under shaded conditions, when the stem fails to harden and remains succulent, the galls are numerous. It is also caused by the fungus *Protomyces macrosporus* (Fig. 3.55).

Fig. 3.55: Symptoms of stem gall on coriander stem.

3.2.25. Black Knot

Black knot is a serious and progressive fungal disease of wild and cultivated *Prunus* species. It is caused by the fungus *Apiosporina morbosa*. The fungus initially infects twigs and branches, causing light brown swelling that turns velvety olive-green. As the disease progresses, these swelling forms hard, rough, coal-black knots or galls that girdle and kill the affected parts. Older knots are often riddled with insects, and a severe infection can stunt and kill the tree (Fig. 3.56).

Fig. 3.56: Symptoms of black knots on prunus trees.

3.2.26. Fruit Spot

Initially the lesions on the fruit are small but conspicuous dark brown, circular which expand as unequal irregular large spots depending on the pathogen involved in the invasion. The appearance of the spot also depends on the genus of the fungal pathogen involved. These spots covered a considerable rind surface area of the fruit giving these a blemished appearance (Fig. 3.57).

Fig. 3.57: Symptoms of fruit spot on pomegranate.

3.2.27. Fruit Rot

Ripe fruits are more susceptibile to the fruit rot fungal infection than green ones. A translucent, water-soaked area appears on the skin of the fruit, which expands in size with the presence of fungal growth on it. Severe infection results in the shriveling and drying of fruits on the fruit-bearing twigs.

When the infection develops after the harvesting of the fruit or in transit and storage, the infected area turns pulpy and covers with the mycelial growth. In such cases, the watery material drains out from the infected area. e.g. fruit rot of citrus caused by *Penicillium expansum.*

The fungal genus responsible for fruit rot are *Penicillium, Phytophthora, Botrytis, Rhizoctonia, Aspergillus, Colletotrichum, Alternaria,* etc. (Fig. 3.58).

Fig. 3.58: Symptoms of fruit rot on citrus.

3.2.28. Grain Mold

The initial symptoms of grain mold are discoloration of grains due to infection and colonization by mold fungi. Grain discoloration varies from light whitish, pinkish, greyish, to shiny black depending on infection and colonization by individual fungal species. Often, grains are colonized by multiple fungi. In severe cases, grains turn completely black. In cases of severe infection at anthesis, grain development is affected, resulting in chaffy florets or small grains on the panicle. If rain occurs during the flowering and grain-filling stages, severe grain molding can occur.

Fungi from many genera have been isolated from the infected sorghum grains and the most frequently occurring genera are *Fusarium, Curvularia, Alternaria*, *Aspergillus,* and *Phoma.*

Fusarium semitectum and *F. moniliforme* develop a fluffy white or pinkish colouration while *Curvularia lunata* colors the grain black. Symptom varies depending on the fungus genera involved and the degree of infection (Fig. 3.59).

Fig. 3.59: Symptoms of grain mould on sorghum earhead.

4

Identification of Fungi Causing Soil-Borne Diseases

4.1. Damping Off

4.1.1. Pythium

Pythium has filamentous (thread-like), hyaline, and coenocytic (non-septate) mycelial growth. The asexual or vegetative stage of *Pythium* produces sporangia (that germinate directly to produce a hypha or indirectly to give rise to a vesicle outside the sporangium, within which zoospores are formed) and sometimes hyphal swellings (spherical sporangia-like structures that do not give rise to zoospores). The vegetative stage of *Pythium* also produces chlamydospores (thick-walled resting spores). The presence of sexual organs like antheridium and oogonium can also be observed in mycelial growth.

The sporangia of *Pythium* are usually produced in water and are non-caducous. Some species have papillate sporangia, but they lack the apical thickening characteristic of papillate sporangia in *Phytophthora*. The protoplasm of the sporangium is discharged through an exit tube to form a vesicle at the end, where zoospores are differentiated and released upon the rupture of the vesicle. Morphologically similar to *Phytophthora*, zoospores of *Pythium* are laterally biflagellate and encyst after a period of motility to germinate into one or more germ tubes that will develop into mycelia. The sporangia can be filamentous, undistinguished from the rest of the mycelium, slightly inflated to form dendroid structures, or greatly inflated to form lobulate or toruloid complexes. The sporangia can be globular, sub-globular, to ovoid, with or without internal proliferation; some may have apical papilla but lack the apical thickening characteristic of those species of *Phytophthora* producing papillate or semi-papillate sporangia. Sexual reproduction occurs by means of oogonia and antheridia. The oogonia can be globular to limoniform, terminal, or intercalary on undifferentiated hyphae with the oogonial wall smooth or ornamented with projections of various lengths and evenly or unevenly distributed. The unicellular antheridia can be sessile or intercalary on the hyphae or terminal on a special stalk; monoclinous if they originate from the oogonial stalk; diclinous if they originate from a different hypha not closely connected with the one subtending the oogonium; or hypogynous if the proximal portion of the oogonial stalk becomes an antheridium.

The oogonium usually contains one or occasionally more than one spherical, smooth oospore which can be aplerotc or plerotic with the entire oogonial space fully occupied by the oospore. The oospore wall can be thin or thick. All species of *Pythium*, with the exception of about seven species (*Py. catenulatum, Py. flevoense, Py. heterothallicum, Py. intermedium, Py. macrosporum, Py. splendens and Py. sylvaticum*) are homothallic, producing sex organs in single cultures.

Hyphal swellings can be present, often indistinguishable from young spherical oogonia or sporangia, and are capable of germination to form a new mycelium; terminal, intercalary, or catenulate, non-deciduous but can be deciduous in *P. intermedium.*

Thick-walled chlamydospores are rare, occurring only in *P. dimorphum* and *P. tracheiphilum*. Appressoria are slight swellings formed by hyphae, especially when in contact with the walls of the Petri dishes, with various shapes and sizes, ranging from sickle-, club- or sausage-shaped to sub-globose single, terminal, or intercalary-forming chains or clusters.

During sexual reproduction, an antheridium fertilizes an oogonium to produce a thick-walled oospore. Some *Pythium* species are heterothallic and require opposite mating types to reproduce sexually but most are homothallic and do not require an opposite mating type (Fig. 4.1).

Standard keys differentiate species within *Pythium*-based on the shapes, sizes, and locations of the sporangium, antheridium, oogonium, and oospore on the mycelium.

4.1.1.1. Pythium aphanidermatum

Colonies on cornmeal agar have a cottony aerial mycelium. On potato-carrot agar, it develops some loose aerial mycelia without a special pattern, and on potato dextrose agar (PDA) it develops heavy aerial mycelia.

Main hyphae are up to 10 µm wide. Zoosporangia consist of terminal complexes of swollen hyphal branches of varying length and up to 22 µm wide. Inside sporangia, asexual spores are present which are lobate (inflated). Zoospores are formed at 15–30° C. Encysted zoospores have a diameter of 12 µm.

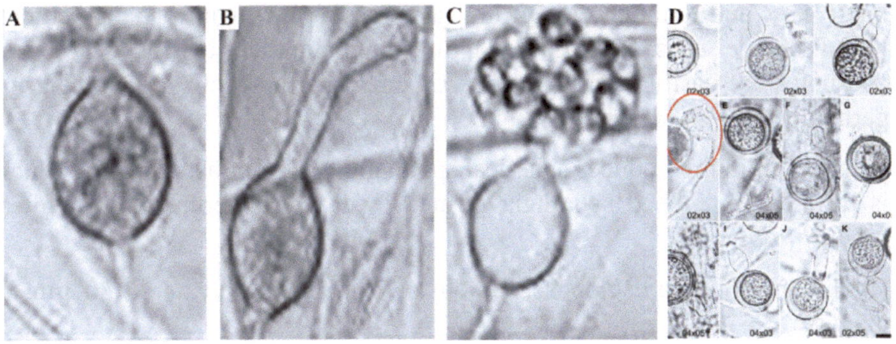

Fig. 4.1: Asexual reproduction of Pythium: **A.** Sporangium; **B.** Direct germination of sporangium; **C.** Formation of zoospore in vesical borne on sporangium; **D.** Sexual reproduction organs *i.e* antheridium and oogonium.

Oogonia are terminal, globose, smooth, and of 20–26 μm (average 24 μm) diameter. Antheridia are mostly intercalary, sometimes terminal, broadly sac-shaped, 11–15 μm long and 9–15 μm wide, 1-2 per oogonium that can be monoclinous or diclinous. Oospores are aplerotic, 20–23 μm (average 21 μm) in diameter and their walls are 1–2 μm thick.

4.1.1.2. Pythium diclinum

Colonies of *Pythium diclinum* on cornmeal agar and potato-carrot agar are submerged and have a more or less radiating pattern.

Main hyphae are up to 6 μm wide. Appressoria are club-shaped. Sporangia are filamentous, non-inflated, branched, or unbranched. At 5° C the fungi form vesicles containing zoospores. Encysted zoospores are 6–7 μm in diameter.

Oogonia are spherical or ovoid, smooth, mostly terminal or subterminal, occasionally intercalary, 19–23 (average 20.5) μm in diameter. Antheridia are typically diclinous, 1–2 per oogonium, about 12 × 5 μm in diameter. Antheridial stalks are not branched. Oospores are single, aplerotic, 17–19 (average 17.5) μm in diameter, with up to 3 μm thick wall.

4.1.1.3. Pythium ultimum

Colonies on Bacto-CMA (Bacto-cornmeal agar) are with aerial mycelium.

The main hyphae are up to 11 mm wide. Appressoria are sickle-shaped. Sporangia and zoospores are readily formed; sporangia are sub-globose, intercalary, and terminal, 26–32 µm (average 28 µm), with one, sometimes two discharge tubes, 23–40 × 3–11 µm, often slightly wider near the base; zoospores are easily released at 20° C; direct germination of sporangia by germ tubes is also present.

Pythium ultimum produces spherical sporangia that germinate directly, which is distinct from lobed sporangia formed by P. aphanidermatum that often produce zoospores.

Hyphal swellings are globose, lemon-form, terminal and up to 30 µm in diameter.

Oogonia are usually terminal, sometimes intercalary, globose, and smooth, 18–25 µm (average 23.5 µm) in diameter. Antheridia are 1–2 per oogonium, sac-like, sessile, mostly monoclinous, and sometimes hypogynous. Oospores are aplerotic, sometimes plerotic, globose, 18–24 µm (average 21 µm) in diameter, with up to 2.0 µm thick wall.

4.1.1.4. Pythium irregulare

Hyphae are typically 5 µm in diameter and can have limoniform swellings. Sporangia are spherical, can be terminal or intercalary and produce zoospores which are about 7–10 µm when encysted.

The antheridia are monoclinous or hypogynous, stalked, 1–3 per oogonium, and can be branched. Oogonia are spherical, smooth-walled, or may have several irregular projections and can be intercalary or terminal. Oospores are mostly aplerotic.

4.1.2. Rhizoctonia

The mycelial colonies are white and cream in color. The young colonies are white, which later on turns to brown as they age. Colony color differences are caused by differences in pigments produced by pathogens in the media.

The hyphae are like a long tube with a septum. The branches of hyphae are tapered and at perpendicular angles. At the point of the branch, there is a curve, slightly narrowed, with a septum. Adult hyphae become stiff because of the thickened cell walls.

The sclerotia are classified as macro-sclerotia and micro-sclerotia. Most macrosclerotia are blackish-brown, small, round-shaped with a size of 1 mm or less. Microsclerotia are shaped like a thin crust on PDA media, originally white and small. The sclerotial color may be white, pale brown, or dark brown and is formed by different distribution patterns of sclerotial bodies on the fungal growth. The type species is *Rhizoctonia solani* (Fig. 4.2).

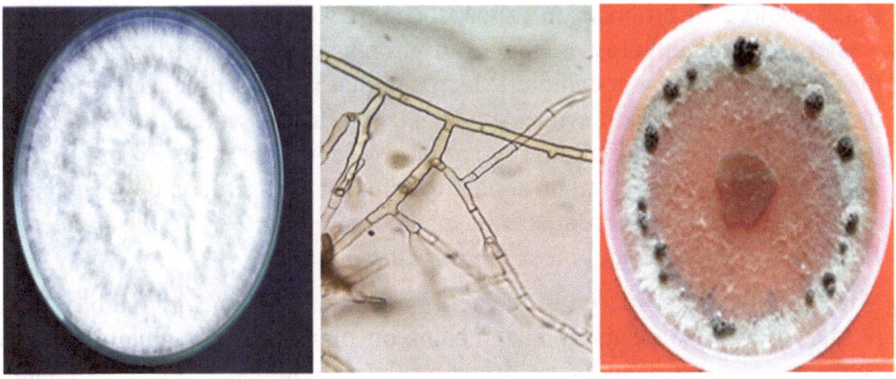

Fig. 4.2: Mycelial growth of Rhizoctonia on PDA, mycelial branching and sclerotial bodies.

4.1.3. Fusarium

Colony morphology of the fungal mycelium varies widely. Mycelia may be floccose, sparse, or abundant and range in color from white to pale violet (Fig. 4.3).

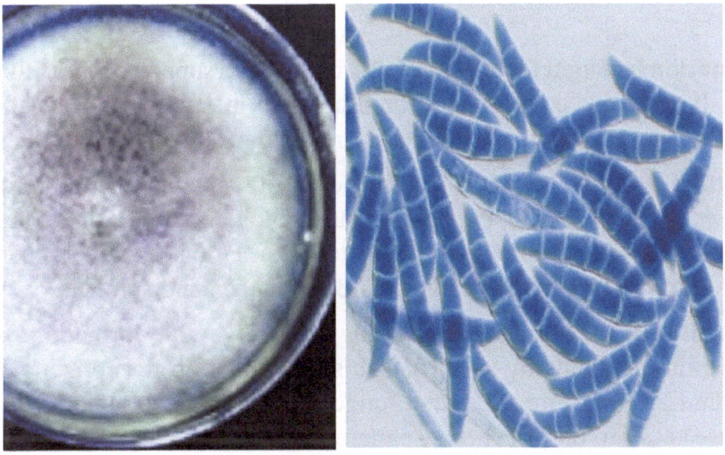

Fig. 4.3: Fungal colony of *Fusarium* and sickle shaped macroconidia.

Fusarium spp. reproduces asexually and produces three kinds of fungal spores known as macroconidia, microconidia, and chlamydospores. Some species of *Fusarium* produce all three types of spores, while others produce them singularly.

These spores, especially the microconidia, are held by microconidiophores.

These conidiophores may be either mono-phialides only or both mono-phialides and poly-phialides in a given species producing microconidia.

Macroconidia are produced in a sporodochium, which is an erumpent crowded cluster of conidiophores arising from stroma to form a cushion-like mass that supports the macroconidia. Macroconidia are also produced on mono-phialides (a conidiophore with a single opening through which an endoconidia is released) and poly-phialides (two or more openings or pores from which the endoconidia are forced out) on aerial mycelium. The macroconidia vary in size and shape. The Major producers of macroconidia are *Fusarium semitectum, Fusarium avenaceum* and *Fusarium suglutinans*.

Microconidia are produced in the aerial mycelium. The microconidia can be produced on false heads or false chains on mono- or poly-phialides. False heads are a result of moisture drops on the conidiophore, and they contain the endoconidia as they are produced. Microconidia have different shapes and sizes; the microconidia produced in chains have a truncate base.

Chlamydospores are thick-walled spores filled with lipid-like material that carries the spores overwinter in the soil. Chlamydospores are sometimes airborne occurring in pairs, in clumps, or chains. They have an outer wall which can be smooth or rough.

The growth of the mycelium and conidia depends on the growth medium of the fungus.

On carnation leaf agar: It promotes sporulation and suppresses mycelial growth. It produces conidia and conidiophores in large numbers, and specialized morphologies of the spores are distinct.

On potato dextrose agar: This is the most valuable medium for *Fusarium* growth, producing gross morphological appearance and colony colorations. The medium promotes sporulation, however, takes longer to grow. The conidia produced are misshapen and atypical.

KCl medium: KCI medium is used to observe the formation of microconidia in chains by species. The species that do form chains of microconidia form more abundant, longer chains on this medium. The chains are easier to observe because there is less moisture on the surface of the agar and fewer droplets of moisture in the aerial mycelium.

Soil agar: Soil agar promotes rapid chlamydospore formation in several *Fusarium* species. A large inoculum with actively growing fusarium inoculates produces chlamydospores within 3–4 days but secondary inoculates produce chlamydospores in 30 days.

4.1.3.1. Fusarium oxysporum

Macroconidia are formed in pale orange color, usually, abundant sporodochia are present. The macroconidia are short to medium in length, falcate to almost straight, thin-walled, and usually 3-septate. The apical cell is short and is slightly hooked in some isolates. The basal cell is notched or foot-shaped. Macroconidia are formed from monophialides on branched conidiophores in sporodochia and to a lesser extent from monophialides on hyphae.

Microconidia usually are 0-septate, may be oval, elliptical, or reniform (kidney-shaped), and are formed abundantly in false heads on short monophialides.

Chlamydospores are formed abundantly in hyphae on the agar surface.

Culture growth is fast on PDA, and prominent mycelia appear within 3 days. Mycelia are characterized as off-white, fairly dense, and showing yellow pigmentation at the periphery. The growth pattern of mycelia is in thick concentric rings. Some cultures have bright pink pigmentation at the centers; some have grey to black, while others have creamish-yellow centers. On the reverse side of the plate, cultures are pink to dark brown in the central portion and shades of dark yellow to dull brown as the mycelia radiates to the periphery.

Mycelia may be floccose, sparse, or abundant and range in color from white to pale violet. Abundant pale orange or pale violet macroconidia are produced in a central spore mass in some isolates.

Small pale brown, blue to blue-black, or violet sclerotia may be produced abundantly by some isolates.

Fusarium oxysporum usually produces a pale to dark violet or dark magenta pigment in the agar but some isolates produce no pigment at all.

4.1.3.2. Fusarium solani

Fusarium solani produces asexual spores (microconidia and macroconidia). Its sexual state is Nectria haematococca (ascomycete). It produces chlamydospores and overwinters as mycelium or spores in infected or dead tissues.

Some isolates are homothallic and may produce red or orange perithecia.

Chlamydospores often are produced abundantly in pairs in hyphae and in the agar.

On potato dextrose agar medium, *Fusarium solani* produces sparse to abundant, white creamy mycelium. Colonies are fast-growing, variable in color and texture, often granular or fluffy, rose-red, purple, or lavender, but may start as white cottony colonies that darken with maturity. The hyphae branch at a 90-angle constriction of hyphae formation of septa at a short distance from the point of the hyphal branch origins and the absence of clamp connection, conidia, and rhizomorphs are characteristics of *F. solani*.

Many isolates do not produce pigments in the agar although some violet or brown pigments may be observed.

Sporodochia are cream, blue, or green on carnation leaf agar and contain numerous macroconidia. Cream sporodochia usually contain more macroconidia than blue or green ones.

Macroconidia are relatively wide, straight, stout, and robust. The apical cell is blunt and rounded. Basal cells may have a distinct foot shape or be poorly developed, ranging from straight to almost cylindrical, usually with a notched or rounded end. Macroconidia are 5–7 septate and usually abundant in sporodochia.

Microconidia are oval, ellipsoid, reniform, and fusiform in shape with zero or one to occasionally two septa and abundant in the aerial mycelia. Aerial mycelium is present as false heads. Microconidia are formed in false heads on very long monophialides. Conidiogenous cells are monophialides, often quite long.

Chlamydospores are commonly formed abundantly and rapidly, usually within 2–4 weeks on CLA. They are present intercalary in the hyphae or formed terminally on short lateral branches usually singly or in pairs, but occasionally in short chains. It appears as globose to oval in shape and smooth or rough-walled.

4.1.3.3. Fusarium equiseti

Fusarium equiseti (Corda) Sac is the anamorph stage while *Gibberella intricans* is the teleomorph of the fungi.

On potato dextrose agar, the mycelial growth is abundant and initially white but later turns brown with age. Sporodochia may form in a central spore mass but may not be obvious since they can be obscured by the mycelium. The spore mass may be pale orange to dark brown, and annular zonations may develop in response to a light-dark cycle. This species forms a pale brown to dark brown pigment where the colony contacts the agar. Dark brown spots or flecks of pigment are usually formed in the agar.

Sporodochia are orange in color. They are present on carnation leaf pieces in CLA. The macroconidia from the sporodochia usually are uniform in shape and size and mainly used for identification purposes of this fungus.

Abundant macroconidia form in orange sporodochia. The thick-walled macroconidia have strong dorsiventral curvature and usually are 5–7-septate, with a distinctly foot-shaped basal cell and a tapering apical cell that may be filamentous or whip-like. Macroconidial length varies from 25 to >120 µm. Macroconidia are produced from monophialides on branched conidiophores in sporodochia. No microconidia are produced. Chlamydospores are produced in hyphae, most often intercalary, solitary, in pairs, frequently forming chains or clusters, and globose (7–13 µm). Chlamydospores are formed relatively rapidly on CLA. The chlamydospores become verrucose and pale brown with age and may give a CLA culture a golden or brown appearance.

4.1.3.4. Fusarium acuminatum

Fusarium acuminatum is a relatively slow-growing species that produces white mycelium, which is abundant in some isolates. The mycelium is floccose with rose to burgundy pigmentation that can be grayish-rose at the periphery. Sporodochia form in the center of the colony in a small central spore mass and are pale orange to dark brown. Red or brown pigmentation is seen on the PDA media. They are found on carnation leaf pieces in CLA.

Macroconidia are formed in pale orange sporodochia. Macroconidia is with moderate equal curvature and thick walls. Basal cell is distinct foot shape, but not as prominent as in species such as *F. equiseti* or *F. armeniacum*. The apical cell tapers and can be relatively elongate. There is often a distinct curve to the latter half of the apical cell. They are typically five-septate, but sometimes 3- and 4-septate too. They are very abundant in sporodochia.

Microconidia are usually absent, but may be formed sparsely by some isolates. When formed, the microconidia are zero or one-septate. They are fusiform or reniform in shape.

Chlamydospore formation is very slow in most isolates and may require more than six weeks. They are formed in hyphae on or in the agar medium. They may be formed in chains or in clusters. Their presence is not a reliable taxonomic criterion.

4.1.3.5. Fusarium verticillioides

On potato dextrose agar medium, initially cultures have white mycelia but may develop violet pigments with age. Pigmentation in the agar varies, ranging from no pigmentation or grayish orange to violet grey, dark violet or dark magenta or almost black.

Macroconidia are formed in pale orange sporodochia that may be rare or can be obscured by thick mycelium or numerous microconidial chains. Macroconidia are relatively long and slender, slightly falcate or straight, and thin walled. Apical cell is curved and often tapered to a point. Basal cell is notched or foot shaped. Macroconidia are 3 to 5-septate. Abundance of macroconidia varies by strain, but macroconidia may be rare in some isolates.

Microconidia are oval to club-shaped with a flattened base and 0-septate. They are always formed from monophialides, which may occur in V-shaped pairs to give a "rabbit ear" appearance, and are found in chains that may be quite long.

Chlamydospores are absent, but swollen cells in the hyphae may be mistaken for chlamydospores or pseudo chlamydospores.

Blue–black sclerotia may develop in some isolates, but are not diagnostic although they may be indicative of a high level of female fertility.

4.1.4. *Phytophthora*

Phytophthora species are identified on the basis of colony morphology, mycelial characteristics, cardinal growth temperatures, morphology, and dimensions of sporangia, oogonia, and antheridia. Morphological differences between some of the species are few and variable, making it difficult to classify the species accurately. More than 50 species are identified based on morphological characteristics (Figs. 4.4–4.6).

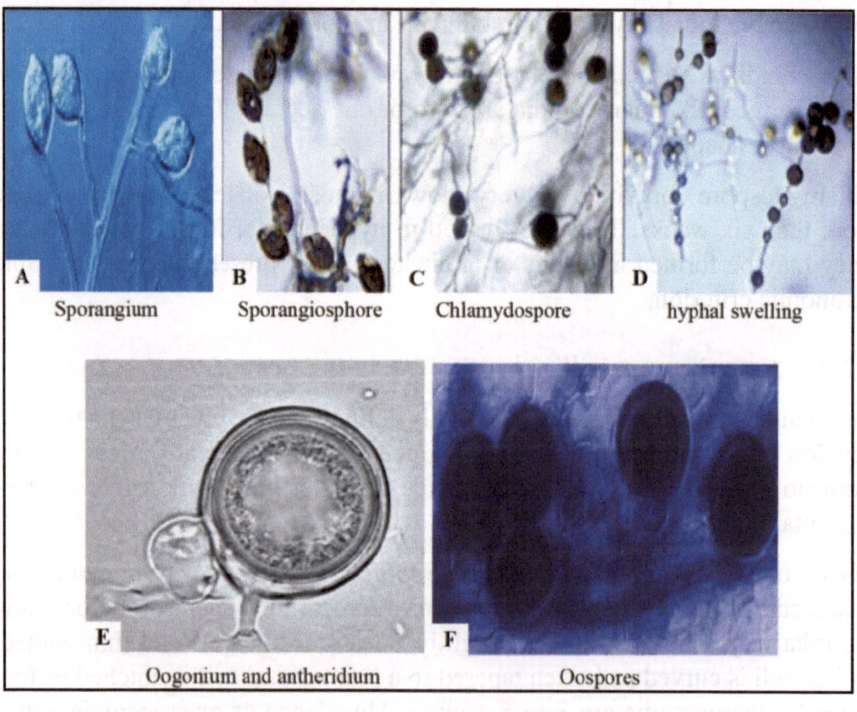

Fig. 4.4: Morphological characters of *Phytophthora* useful for its identification.

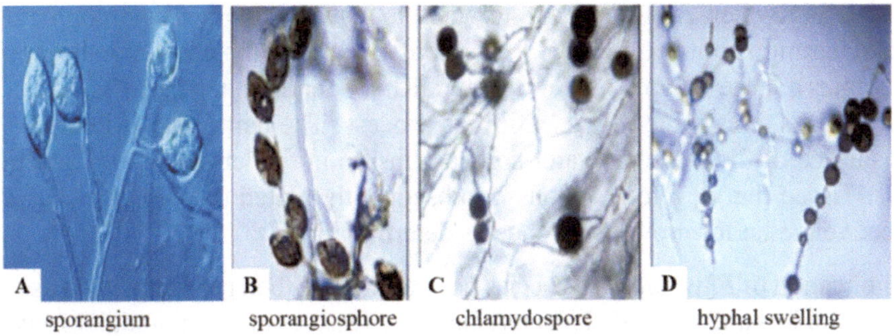

Fig. 4.5: Asexual phases observed in *Phytophthora* sp.

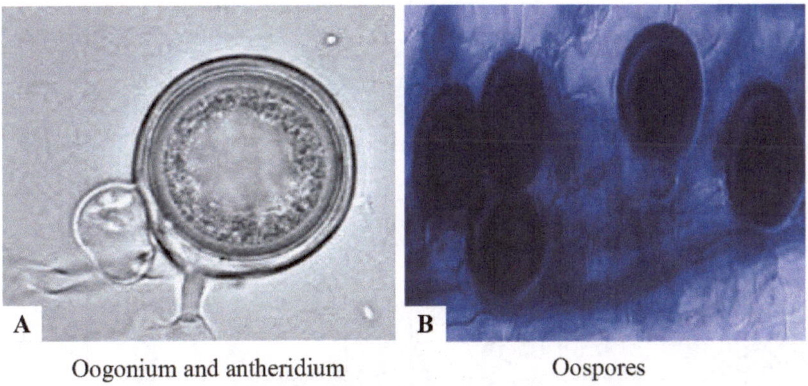

Fig. 4.6: Sexual phase. **(A)** Sex organs oogonium and antheridium; **(B)** Oospores.

Microscopic observation of tissue sections may reveal the presence of oospores and coenocytic hyphae.

The fungus *Phytophthora* and its species can be detected by plating diseased tissue onto a selective medium to allow hyphae to grow from the infected plant material when plates are stored in the dark at 20° C. The most commonly used selective media have corn meal or V8-juice bases and incorporate ingredients like pimaricin, ampicillin, rifampicin, and PCNB. Hymexazol may be added for better inhibition of *Pythium* species, but it also restricts the growth of many species of *Phytophthora*.

Another method that can be useful in the isolation of *Phytophthora*, particularly when extensive decay and deterioration of tissues have occurred, is the baiting of soil from diseased plants with pears and other fruits or various leaf tissues, which are known to support the growth of a wide range of *Phytophthora* species. Baiting involves the use of plant material to recover *Phytophthora* spp.

from soil or water. Zoospores are attracted to leaves and other common baits. While baiting is not quantitative, it is a sensitive method of detection because it samples a large volume of soil or water. Rhododendron and camellia leaves, or leaf disks, are commonly used in this process. The bait tissue is floated on the surface of the water or water-soil suspension in the laboratory for several days. Tissue that develops lesions is subsequently plated onto *Phytophthora*-selective media for growth and species determination.

Identification of *Phytophthora* species has traditionally been accomplished using morphological and cultural characteristics. The features of the sporangium, such as the shape and the type of a papilla, are important in identification. Sporangia may be classified as papillate, semi-papillate, or non-papillate (Figs. 4.7 and 4.8).

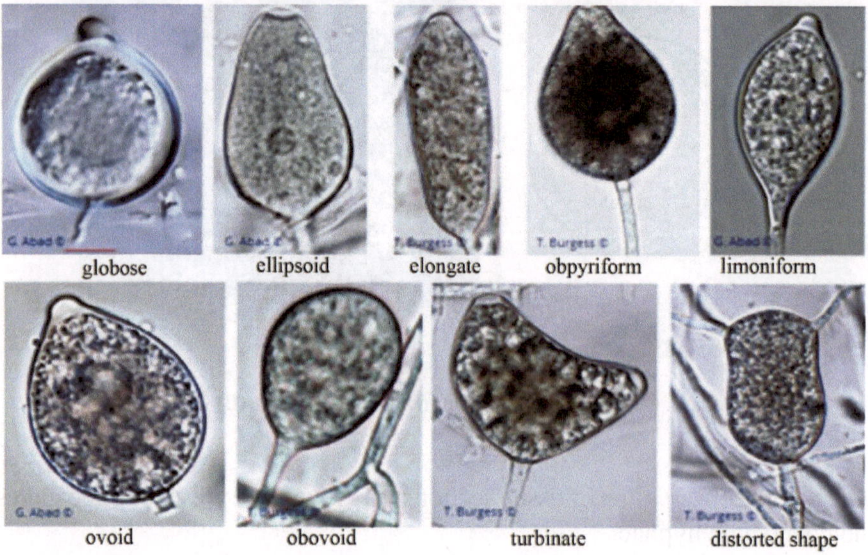

Fig. 4.7: Shapes of sporangia in *phytophthora.*

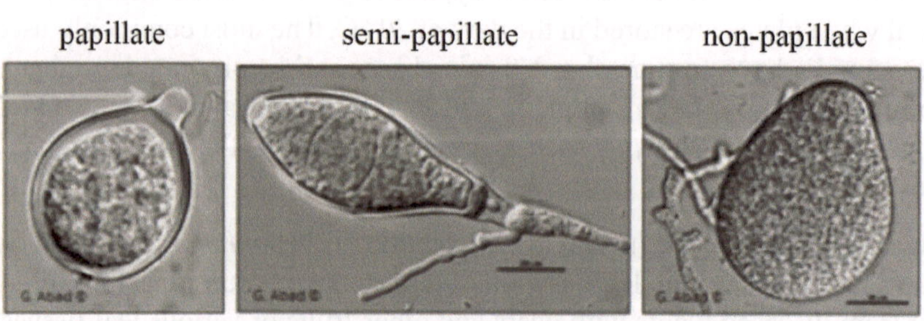

Fig. 4.8: Sporangial papillation in *Phytophthora.*

The caducous or deciduous nature of sporangia and the subsequent length of the subtending pedicel are key features of some species of *Phytophthora*. The homothallic or heterothallic nature and the resultant production or lack of production of oospores are often important identifying attribute. Oospore morphology can also be an important feature for some species.

Many species of *Phytophthora* require flooding with soil extract or distilled water under light to induce sporangium formation. Culture plates can be flooded directly, or agar plugs from the colony margin can be taken and flooded within a smaller (60 mm diameter) petri plate. If no sporangia form within 24 h, a brief chilling period followed by the return to room temperature may induce sporangium formation.

Phytophthora species may reproduce sexually or asexually. In many species, sexual structures have never been observed or have only been observed in laboratory matings. In homothallic species, sexual structures occur in a single culture. Heterothallic species have mating strains designated as A1 and A2. When mated, antheridia introduce gametes into oogonia, either by the oogonium passing through the antheridium (amphigyny) or by the antheridium attaching to the proximal (lower) half of the oogonium (paragyny), and the union produces oospores.

Asexual (mitotic) spore types are chlamydospores and sporangia, which produce zoospores. Chlamydospores are usually spherical and pigmented, and they may have a thickened cell wall to aid in their role as a survival structure. Sporangia may be retained by the subtending hyphae (noncaducous) or shed readily by wind or water tension (caducous), acting as dispersal structures. Also, sporangia may release zoospores, which have two unlike flagella, which they use to swim toward a host plant. Whereas fungal cell walls are made primarily of chitin, *Phytophthora* cell walls are constructed mostly of cellulose.

Species of *Phytophthora* can be readily identified by the aseptate hyphae with slight constriction at the base of its initially right-angled branches; by the ovoid, obpyriform to limoniform sporangia produced in succession on a typically sympodially branched, undifferentiated sporangiophore of indeterminate growth; by the differentiation of mature, laterally biflagellate zoospores within the sporangium proper; by its globose oogonium with a single spherical oospore with thin or no periplasm; and by its antheridial configuration being amphigynous and paragynous. The taxonomy of *Phytophthora* species has been traditionally based on morphological characters, e.g. the papillation, size, length and breadth ratio, and caducity of sporangia as well as the ability to produce oospores in single cultures, antheridial configuration, and oogonial size.

Hyphal swellings are not cut off by septa and, when present, may be single or clustered, terminal, intercalary, or branching to form a network; spherical, oval or irregular; and vary considerably in size. They may be formed on agar plates, but more commonly in water. Chlamydospores are sometimes confused with spherical hyphal swellings, but they can be distinguished by a secondary wall, the presence of a delimiting septum, and usually the internal reorganization of the cytoplasm. They are hyaline to yellowish or brownish, thin to thick-walled (mostly 1–2μm), terminal or intercalary (Figs. 4.9 and 4.10).

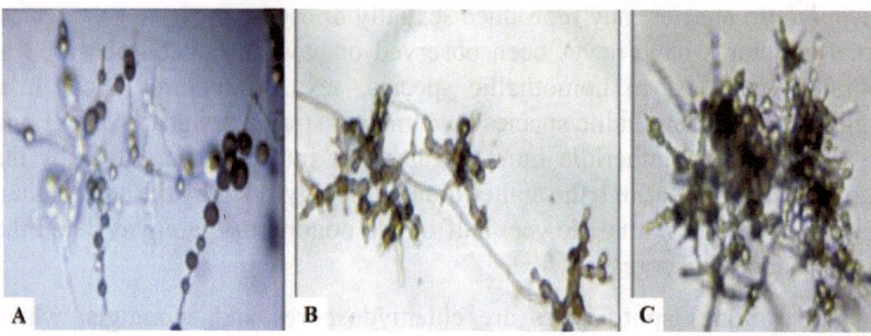

Fig. 4.9: Hyphal swellings. **(A)** globose, sub-globose, catenulated; **(B)** lobate, irregular, radiating; **(C)** irregular, radiating, small, clustered.

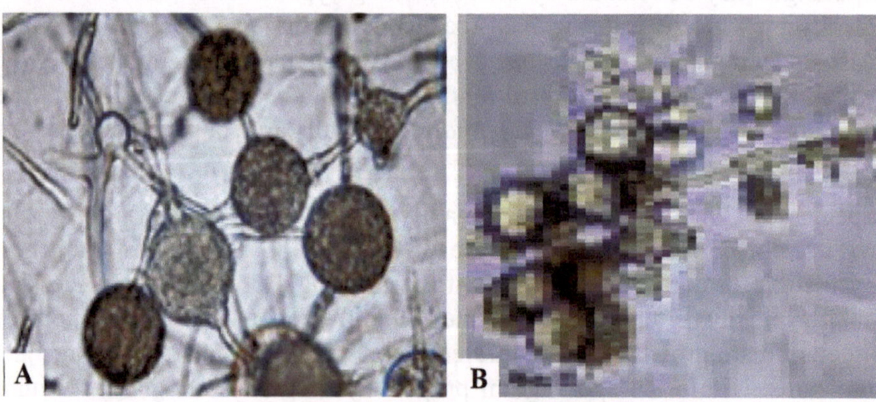

Fig. 4.10: Chlamydospores. **(A)** globose, sub-globose, irregular, radiating; **(B)** globose, irregular, clustered, radiating.

There are three basic types of sporangia, based on the characteristics of the apex: (a) conspicuously papillate with hemispherical apical thickening over 3.5 μm deep; (b) inconspicuously papillate (semi-papillate) with shallow apical thickening under 3.5 μm deep; and (c) non-papillate, without noticeable apical thickening. A plug may be formed in the sporangiophore at the base of a papillate or semi-papillate sporangium. Just prior to zoospore release or when the non-papillae sporangia are exposed to air, a slight thickening might be detected in non-papillate sporangia which thus appear as "semi-papillate." The papillate and semi-papillate sporangia, may be formed on agar media or host tissues, non-deciduous or deciduous in water with pedicels of variable length, and have narrow exit pores (under 7 μm) after the dissolution of the apical thickening. The non-papillate sporangia are usually formed only in water, are non-deciduous, have wide exit pores (often over 10 μm), and the base of sporangium is not plugged so that the sporangiophore is capable of internal proliferation by producing a new sporangium "nested" within or extended on a new sporangiophore beyond the exit pore of the empty sporangium which often collapses after zoospore dehiscence. The process of internal proliferation can be repeated several times. Laterally, biflagellate zoospores are formed within the sporangium and released in an evanescent vesicle, with the dissolution of the apical thickening for papillate and semi-papillate sporangium or inflation of the apex for non-papillate sporangium.

The zoospore has two flagella inserted in the middle of a longitudinal groove of the ovoid zoospore (10–17 × 7–12 μm), with the anterior end slightly tapering while the posterior end is roundish. The zoospore shape varies depending on the angle of observation, appearing reniform or kidney-shaped only when viewed from the posterior end of the zoospore. The shorter anterior flagellum propels the zoospore with the longer, posterior flagellum trailing behind, serving as the rudder, and it glides smoothly in a spiral fashion while rotating clockwise along its longitudinal axis. In water, after a period of motility that can last up to 24 h or more, the naked zoospore rounds up to form a spherical cyst (7–14 μm diameter) with a cell wall, while the flagella coil up and are discarded. An encysted zoospore may germinate by a germ tube, with or without a germ sporangium, or by another zoospore, a process called "repeated emergence." Occasionally, the sporangium can also germinate directly by one or more than one germ tubes at or near the apex, or close to the sporangiophore. The sporangium has been described as limoniform, obpyriform, ovoid or ellipsoidal, usually terminal, single, basally attached on an undifferentiated, unbranched, or sympodially (close or lax) to irregularly branched, narrow sporangiophore (1–3 μm wide). Occasionally, intercalary sporangia may be produced.

Morphologically, both phytophthora and pythium are very similar in having coenocytic, hyaline, and freely branching mycelia, and oogonia with usually single oospores, but the definitive differentiation between them lies in the mode of zoospore differentiation and discharge. In *Phytophthora*, the zoospores are differentiated within the sporangium proper and, when mature, released in an evanescent vesicle at the sporangial apex, whereas in *Pythium*, the protoplast of a sporangium is transferred, usually through an exit tube, to a thin vesicle outside the sporangium, where zoospores are differentiated and released upon the rupture of the vesicle.

Many species of *Phytophthora* are destructive pathogens of especially dicotyledonous woody trees, shrubs, and herbaceous plants whereas *Pythium* species attacked primarily monocotyledonous herbaceous plants,

Diseases caused by important species of Phytophthora are listed below:

Phytophthora agathidicida causes collar-rot on New Zealand kauri (*Agathis australis*), New Zealand's most voluminous tree, an otherwise successful survivor of the Jurassic.

Phytophthora cactorum causes rhododendron root rot affecting rhododendrons, azaleas, and causes bleeding canker in hardwood trees.

Phytophthora capsici infects *Cucurbitaceae* fruits, such as cucumbers and squash.

Phytophthora cinnamomi causes cinnamon root rot affecting forest and fruit trees, and woody ornamentals including arborvitae, azalea, chamaecyparis, dogwood, forsythia, fraser fir, hemlock, Japanese holly, juniper, pieris, rhododendron, taxus, white pine, American chestnut and Australian woody plants, especially eucalyptus and banksia.

Phytophthora citricola causes root rot and stem cankers in citrus trees.

Phytophthora fragariae causes red root rot affecting strawberries.

Phytophthora infestans causes the serious disease known potato (late) blight: responsible for the Great Famine of Ireland.

Phytophthora kernoviae a pathogen of beech and rhododendron, also occurring on other trees and shrubs including oak, and holm oak. First seen in Cornwall, UK, in 2003.

Phytophthora lateralis causes cedar root disease in Port Orford cedar trees.

Phytophthora megakarya causes cocoa black pod disease and, is invasive and probably responsible for the greatest cocoa crop loss in Africa.

Phytophthora multivora discovered in analysis of isolates with *P. cinnamomi* dieback infections of tuart forests of Southwest Australia, which were previously diagnosed as *P. citricola*. The species was found occurring on many other taxa, so named *multivora*.

Phytophthora nicotianae infects onions and tobacco.

Phytophthora palmivora causes fruit rot in coconuts and betel nuts.

Phytophthora ramorum infects over 60 plant genera and over 100 host species; causes sudden oak death disease.

Phytophthora quercina causes oak death.

Phytophthora sojae causes soybean root rot.

The morphological characteristic of some of the important *phytophthora* species are as follows:

4.1.4.1. Phytophthora parasitica

The hyphae are hyaline and aseptate, sometimes with hyphal swellings. *Phytophthora* parastica produces asexual sporangia, zoospores and chlamydospores. The sporangia are difficult to release from the hyphae, which is different from the, airborne sporangia of *P. infestans that can be released easily*. Zoospores arc produced by the sporangia and are wall-less cells with two flagella, which enable them to swim. Chlamydospores are thick-walled, multinucleated asexual spores, usually produced at the tips or in the middle of the hyphae. *P. parasitica* is predominantly heterothallic, requiring A1 and A2 mating types for the production of oospores. Maximum radial growth of this pathogen is on czapeck-dox agar after eight days of incubation. The optimum temperature for the mycelial growth of the fungus is 20° C followed by 25° C and the least growth is at 40° C.

4.1.4.2. Phytophthora infestans

Phytophthora infestans is a coenocytic oomycete with rare cross walls. Asexual reproduction is *via* sporangia that are ellipsoid to lemon shaped with a small pedicel. Sporangia are 29–36 × 19–22 μm. Sporangia germinate either directly to form a germ tube (at temperatures of 15–24° C), or indirectly *via* zoospores (at temperatures below 18° C). Zoospores (7–12 per sporangium) have two flagella, one forward-directed tinsel type and a backward-directed whiplash type (heterokont). Zoospores are usually uninucleate, but binucleate zoospores have been detected.

In culture, the mycelium is white and fluffy; the colony is somewhat slow growing. Growth rates can vary dramatically among isolates, but fast-growing isolates can cover a 9-cm plate within 7–10 days. Some isolates produce a lumpy appearance: this has sometimes been associated with the A2 mating type. Oogonia rare in host or single culture, but developing promptly on pairing isolates of opposite compatibility type, 38 (max. 50) μm diameter, tapering at base. Antheridia amphigynous, elongated, cylindrical, 17 (max. 22) × 16 μm. Oospores average 30 μm, aplerotic, wall 3–4 μm.

4.1.4.3. Phytophthora megasperma

Colonies and cardinal temperatures

Colony morphology after seven days of growth on potato dextrose agar, V-8 agar, and malt extract agar media are with no distinct pattern. Minimum growth temperature for this fungus is 3° C, optimum growth temperature 18° C, and maximum growth temperature 27° C.

Asexual phase

Sporangia are non-papillate, persistent, elongate, obpyriform, limoniform, or distorted shapes (48–99 L × 29–46 W μm), some with tapered base; with external and internal extended and nested proliferation; originated in unbranched sporangiophores. Hyphal swellings are globose, sub-globose to elongate with radiating hyphae. Chlamydospores are absent.

Sexual phase

For Homothallic fungi oogonia are spherical (16–61 μm diameters). Antheridia are spherical or ellipsoid (14–20 × 10–18 μm), predominantly paragynous and oospores are aplerotic (11–54 μm diameters).

Most typical characteristics

Characterization of Phytophthora megasperma is by the presence of non-papillate sporangia with internal extended and nested proliferation and the predominantly paragynous antheridia, as well as by the shape of hyphal swellings.

4.1.4.4. Phytophthora drechsleri

Colonies and cardinal temperatures

Colonies grown on V-8 Agar, PDA, and MEA media are with no distinctive pattern. On PDA and MEA media the fungus has very light chrysanthemum pattern. Minimum temperature for growth is 3° C, optimum temperature range 27–33° C, and maximum temperature 36° C.

Asexual phase

Sporangia are non-papillate, persistent, ellipsoid, ovoid, irregular shapes, and some with tapered bases (23–34 × 36–59 μm), showing internal and external proliferation, and originated in unbranched sporangiophores. Hyphal swellings are globose, catenulate (rosary shape), and frequently clustered. Chlamydospores are absent.

Sexual phase

Heterothallic fungi. Oogonia are with smooth wall, globose, some with slightly tapered bases (within antheridium), 25–40 μm diameters. Antheridia are amphigynous, oval (9–16 μm). Oospores are almost plerotic to aplerotic (22–35 μm diameters) with thin walls (2.5 μm).

Most typical characteristics

Characterization of *Phytophthora drechsleri* is by the presence of typical hyphal swellings that are globose with rosary shape, frequently clustered, and produced in water cultures.

4.2. Root Rot

4.2.1 Rhizoctonia

The mycelium is initially white in color which later turns dark brown to black as and when sclerotial formation starts. The vegetative mycelium has the formation of septum, or constriction, near the junction of branching. Branching occurs mostly at right angles to parent hyphae.

Fungal structures, like microsclerotia, are also present. Moniliod cells are produced. Aerial mycelium aggregated to form numerous dark-brown to black-colored sclerotia noticed on the reverse side of the culture plate. The sclerotia vary in shape, size, and growth pattern of scattered to clustered. The type species is *Rhizoctonia bataticola* (Fig. 4.11).

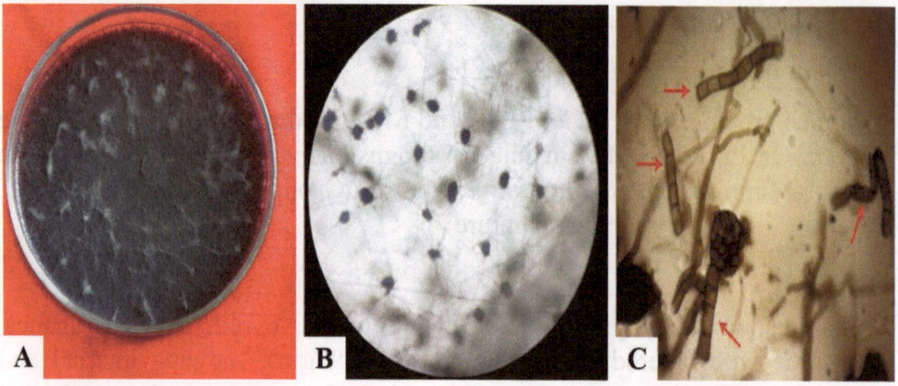

Fig. 4.11: *Rhizoctonia bataticola*; **(A)** pure culture, (B) sclerotial bodies on the reverse side of the culture plate; **(C)** moniliod cell formation (40×) and initial stage of microsclerotia formation.

4.2.2. Sclerotium

The fungus produces white, dense, radial mycelial growth on potato dextrose agar medium. In the early stages, the fungus produces white mycelium gradually loses its luster, and becomes somewhat dull in appearance. Aerial hyphae are not uniformly distributed. Initiation of sclerotial bodies occurs from the fifth day onwards after inoculation. In the beginning, the sclerotial bodies are white which, gradually turns to a buff-brown color and then to chocolate brown at maturity. The fully matured sclerotia are spherical, ellipsoidal, and measured 1.5 mm in diameter (Fig. 4.12).

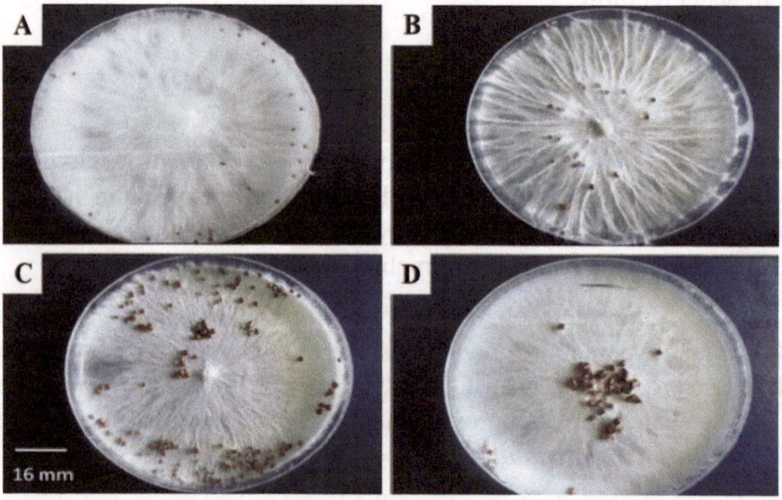

Fig. 4.12: Growth of *Sclerotium rolfsii* on potato dextrose agar media; **(A)** radial growth of mycelium, **(B)** sclerotial initiations, **(C and D)** matured sclerotial bodies.

4.2.3. Phytophthora

Following species of *Phytophthora* are mainly involved in root rot disease.

1. *Phytophthora cactorum*
2. *Phytophthora cinnamomi*
3. *Phytophthora alni*
4. *Phytophthora citrophthora*

Please refer "*Phytophthora*" section for identification characters of *Phytophthora*.

4.2.4. Pythium

Following species of *Pythium* are mainly involved in root rot disease.

1. *Pythium ultimum*

   (Refer *"Pythium ultimum"* section)
2. *Pythium aphanidermatum*

   (Refer "*Pythium aphanidermatum*" section)
3. *Pythium irregulare*

   (Refer "*Pythium irregulare*" section)
4. *Pythium myriotylum*

4.2.5. Fusarium

Following species of *Fusarium* are mainly involved in root rot disease.

4.2.5.1. Fusarium solani f. sp. phaseoli

The growth of aerial mycelia is luxuriant, moderately luxuriant, and scanty. Mycelial texture is either fluffy or fibrous. The mycelial colors are purple, pink and white. The hyphae of the cultured isolates are highly branched, slender, and septate and produce conidia and chlamydospores. The microconidia are 0–1 septate, and the mean microconidial length and width is 9 × 3.5 µm. Macroconidia has 3–7 septation. The mean macroconidial length and width is 36 × 4.0 µm.

4.2.5.2. Fusarium solani f. sp. pisi

This fungus can be grown on potato dextrose agar (PDA), malt extract agar (MEA) and carnation leaf agar (CLA) media in the darkness at 27° C to obtain the fungal growth after seven days.

The aerial mycelial growth is white and cottony on MEA and white to milky on PDA media. Pigmentation ranging from yellowish to orange-brown is present on the reverse side of the plate. Sporodochia emerges after seven days of incubation as white flowers on pieces of carnation leaf placed on CLA. On this aerial mycelium, erect, mononematous conidiophores are abundantly produced from the agar surface. Mononematous conidiophores are acremonium-like and unbranched or occasionally branched once.

Long conidiophores range from 2.1 to 162.2 μm with a mean length of 45.1 μm and 4.8 μm width. Short conidiophores range from 3.0 to 35.2 μm with mean length of 11.2 μm and 4.0 μm width at the base, terminating into a single sub-cylindrical phialide. Tip of the phialide with inconspicuous preclinical thickening, collarette not flared. Macroconidia is abundant, 4.3–46.1 × 5.4–6.2 μm, 2–4 septate, slightly curved, or arcuate with a rounded apical cell and wedge-shaped, weakly pedicellate basal cell. Microconidia is ovoidal or with a rounded apex and truncate base, 2.5–12.7 × 1.5–2.1 μm. Chlamydospores are absent. Cardinal growth temperature is 25–30° C which gives optimal mycelial development of 17–21 mm diameter on PDA and Spezieller Nährstoffarmer agar (SNA) media, respectively. Growth of colonies is absent at 5 and at 35 and 40 °C on PDA.

4.2.6. Rosellinia

Rosellinia produces cobwebby mycelia that may be visible on the lower trunk, in soil over infected roots, growing from infected roots, or beneath bark of the root crown and major roots. These whitish growths of *Rosellinia* are much smaller and more scattered in comparison with the fanlike mycelia of Armillaria root disease found under bark. Mycelia of *Rosellinia* also lack the characteristic mushroom-like odor produced by fresh mycelia of *Armillaria.*

When *Rosellinia* root rot is the cause of disease, exposing roots and removing the cortex (outer layer) of infected roots will reveal a whitish, fluffy layer of fungal mycelia. Older mycelia of *Rosellinia* can become blackish, brown, or tan and may occur crust-like on dead roots or the root crown. Dark cankers may be present in the lower trunk and main roots of infected hosts.

When *Rosellinia*-infected bark or root pieces are sealed in a moist chamber such as a plastic bag or jar for a few days, abundant, white mycelia are produced. Tiny, stalked spore-forming structures (coremia) with fluffy white growths on the end may also be produced and visible with a hand lens (Fig. 4.13).

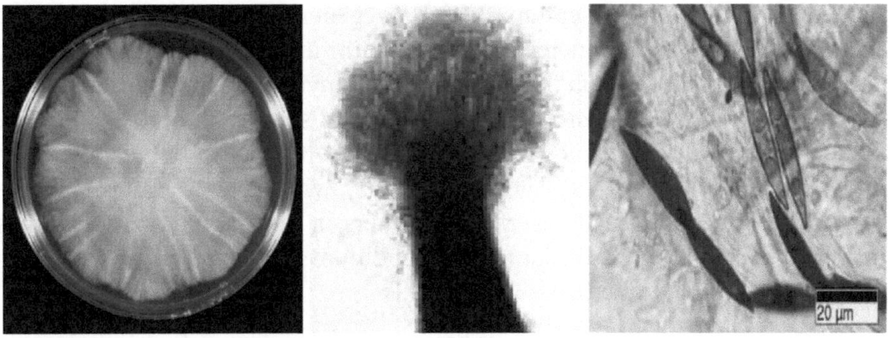

Fig. 4.13: Fungal growth of *Rosellinia,* Synnemata and fungal spores.

Following species of *Rosellinia* are mainly involved in root rot diseases:

4.2.6.1. Rosellinia pepo

A major diagnostic characteristic at the generic level is the presence of pear-shaped swellings at the septa of the hyphae. At the species level, *Rosselina pepo* and *Rosselina bunodes* are distinguished based on the mycelial aggregates formed on the roots. *R. pepo* produces grayish cobweb-like strands, which become black and coalesce into a wooly mass. Under the bark, white, star-like fans are macroscopically visible.

4.2.6.2. Rosellinia bunodes

Specialized structures associated with sexual or asexual reproduction are absent. However, the typical pear-shaped swelling at the septa of mycelia, which is characteristic of the genus *Rosellinia* is present. *Rosellinia bunodes* shows black branching strands firmly attached to the roots, forming black dots and lines embedded in the tissues.

4.2.6.3. Rosellinia necatrix

On culture media, the young mycelium of *Rosellinia necatrix* is initially white and cottony, but with age, it becomes brown–black in color. The coloration is more intense as the malt concentration in the medium increases and less intense with increased light. In general, *R. necatrix* is a fast-growing fungus, and its mycelium covers the surface of the culture medium when it is incubated in the dark at 20–24° C. The most important morphological characteristic of *R. necatrix* hyphae is the existence of pear-shaped or pyriform swellings immediately above the septum, which is generally used to identify the species. They have a diameter of up to 13 μm and are more abundant and larger in superficial mycelia than in submerged mycelia. Under special culture conditions *in vitro*, *R. necatrix* develops mycelial strands, or cords, which are composed of wide filamentous hyphae arranged parallel to one another as a differentiated prosenchyma. They are initially white and weak but become black when old, can reach several centimeters in length, and are sometimes branched and cylindrical. Under natural conditions, they are fungal propagules in soil and on root surfaces.

Sclerotia are black, hard, and spherical nodules, several millimeters in diameter, located mainly on invaded roots and connected from their base with the subcortical mycelium. Black sclerotial crusts formation takes place on roots. When cultured on potato dextrose agar in the light, *R. necatrix* forms microsclerotia, but these are not formed when grown on various synthetic media in darkness. Microsclerotia are formed as irregular bodies, which are rough and composed of a compact mass of interwoven hyphae with no differentiated cells and with raised melanin content.

Rosellinia necatrix produces three different spores such as chlamydospores, conidiospores and ascospores. Chlamydospores can be found under exceptional environmental conditions but are rarely found under natural conditions. They are formed from pyriform swellings in which protoplasm condenses and a wall formation takes place that divides the spherical zone from the remaining hypha. Conidia originate at the end of the synnemata conidiogenous cells. Synnemata is present on plants killed by *R. necatrix*, either at soil level or on the roots. They are formed from sclerotia or from brown mycelial masses and variable in number. The length of the synnemata varies between 0.5 and 1.5 mm, and conidia are 3–5 μm long and 2.5–3 μm wide. Conidial germination is very difficult to induce *in vitro*. *R. necatrix* ascospores are formed inside the perithecium and when they reach maturity, are expelled in a mucilaginous mass from the pore or papilla located on the top of the perithecium.

4.2.7. Phellinus

Phellinus species attacks a wide range of tropical plants, although mostly trees. The leaves of an infected tree show yellowing followed by wilt and typical dieback symptoms. Symptoms may develop slowly or the tree may wilt and become defoliated in only a few days.

The most characteristic symptom of this disease is the brown encrustation covering the surface of the diseased roots. This consists of brown mycelium in which soil and small stones are firmly embedded. The fungus moves toward the collar of the tree and occasionally the encrustation may be visible above ground level. In the diseased wood, dark lines are visible due to the presence of the fungal hyphae. In advanced stages of decay, the wood becomes light, dry and friable and honeycombed. It is one of several fungi associated with heart or butt rots of forest and timber trees. Sporophores are very rare, large, hard purplish-brown bracts with yellowish-white growing margins and concentric blackish zones toward the edges. They are formed above ground on the encrustation on the trunk. Unlike other similar fungi, there are no rhizomorphs. Spread is by physical contact with the root encrustations (Fig. 4.14).

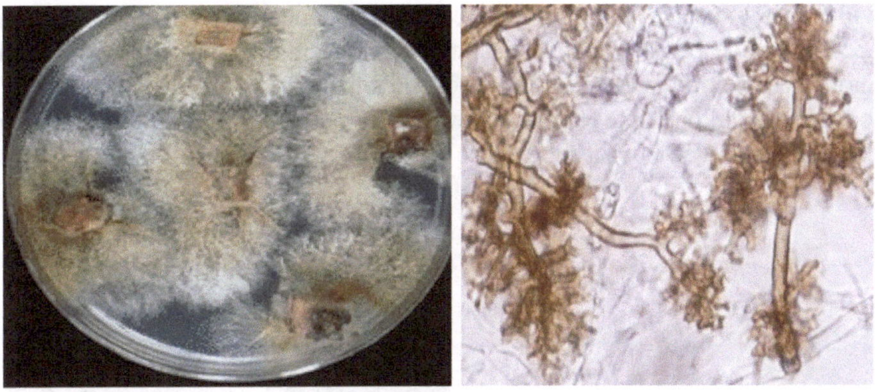

Fig. 4.14: Fungal growth and conidia of *Phellinus.*

Following species of *Phellinus* are mainly involved in root rot diseases

4.2.7.1. Phellinus noxius

Raised white and brown plaques are characteristic of *Phellinus noxius* in malt culture. Colonies on potato dextrose agar (PDA) are white, but with age becoming brown with irregular lines or patches of darker tissue. Hyphae have clamp connections, but they are not commonly produced in culture. Arthrospores and trichocysts are absent in the field, but are produced abundantly in culture.

4.2.7.2. Phellinus weirii

Abundant, characteristically long, brown septal hyphae, 5–10 μm in diameter and up to 3 mm long, with a thickened wall, 1.5–2.5 μm is found between the sheets of decayed wood. Mycelium lacks clamp connections. Basidiospores are globose to sub-globose becoming oblong-ellipsoid, with a small apiculus, smooth and hyaline. Size of basidiospore is 3.6–4.5 × 2.7–3.5 μm.

4.2.8. Gibberella

Gibberella is the sexual stage for the *Fusarium* fungi. The fungus has both asexual (*Fusarium*) and homothallic sexual (*Giberrella*) life cycle stages that can occur together. In the sexual stage, perithecia develop on the mycelia and give rise to ascospores, which are the primary inoculum. Asexual macroconidia may contribute to a primary or secondary infection.

Perithecia are very dark blue to dark purple in color, spheroid to ovoid in shape, with a warty surface and a pronounced ostiole. Generally, perithecia are borne on the surface of the substrate, scattered or in clumps. The asci contain an unordered meiotic tetrad of eight ascospores that can be straight or curved, 1- to 3-septate, and hyaline to pale yellow. *Gibberella coronicola* is the important pathogen involved in root rot disease (Fig. 4.15).

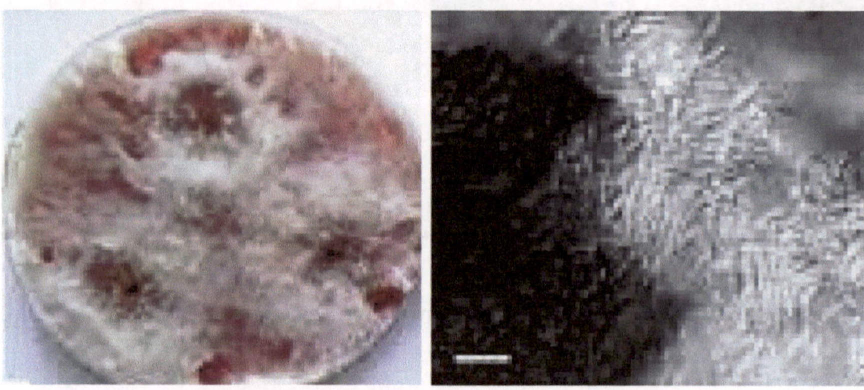

Fig. 4.15: The fungal growth and perithecia of *Gibberella.*

4.2.8.1. Gibberella coronicola

Heterothallic production of perithecia is found. Mature asci and viable ascospores are formed. Perithecia in the fertile pairings are sub-globose to ovoid, dark, 120–370 μm diameter and form directly on the surface of rice stems placed on the culture media. Asci are unitunicate and eight-spored when mature. Mature ascospores are primarily hyaline, fusoid, straight, or curved, with rounded ends and 1–3 septate.

4.2.9. Rhizopus

The pathogen is reported as a cause of root rot in sugar beet. This disease first appears as a temporary wilting of foliage during periods of stress; as the disease advances, the wilting becomes permanent. After the death of the beet, the foliage and root become very brittle and dry. Infected root tissue appears gray-brown with darker vascular rings. The disease generally progresses downward, with the infected tissue becoming dark and spongy. The taproot can eventually be completely consumed with white mycelium. Black sporangia (spores) are produced on the white mycelium, giving the fungal mass a dark appearance. Often, the fungus will decay the internal tissue, creating a cavity filled with clear fluid. The roots may have an odor of acetic acid.

Rhizopus fungi are characterized by a body of branching mycelia composed of three types of hyphae: stolons, rhizoids, and usually unbranching sporangiophores. The black sporangia at the tips of the sporangiophores are rounded and produce numerous non-motile multinucleate spores for asexual reproduction.

The genus Rhizopus is characterized by the presence of stolons and pigmented rhizoids, the formation of sporangiophores, singly or in groups from nodes directly above the rhizoids, and apophysate, columellate, multi-spored, and generally globose sporangia. After spore release the apophyses and columella often collapse to form an umbrella-like structure. Sporangiospores are globose to ovoid, one-celled, hyaline to brown and striate in many species.

The fungal colonies are fast growing and cover an agar surface with a dense cottony growth that is at first white and turns grey or yellowish brown with sporulation. Sabouraud dextrose agar (SDA) with cycloheximide at 37° C produce a characteristic dense cottony-fluffy growth with a flossy texture. The hyphae are non-parallel and branched.

Rhizopus stolonifer and *R. arrhizus* are common in most agricultural soils throughout the world. In spite of their wide distribution, both fungi are weak root rot pathogens and tend to only be a problem when the crop is compromised by some other factor such as excess soil moisture, crown injuries, or insect injury (cutworms, armyworms) to roots. While the symptoms caused by these species are identical, the optimum temperatures for disease development are different: *Rhizopus stolonifer* causes disease at low temperatures of 14–16° C, while high temperatures, 30–40° C, favor *R. arrhizus* (Fig. 4.16).

Fig. 4.16: Structure of *Rhizopus* fungus.

4.2.9.1. Rhizopus stolonifer

Rhizopus stolonifer is also known as black bread mold. Thallus is a white cottony, much-branched mycelium. Mycelium is differentiated into nodes and internodes. The nodal region bears many branched rhizoid that grow downward inside the substratum for anchorage and absorption of food. The internodal region is the aerial and arching hyphae, known as stolons which, when touched by the substratum, form the nodal region.

4.2.9.2. Rhizopus arrhizus

Colonies are fast growing, with some tendency to collapse, white cottony at first that turns brownish grey to blackish-grey depending on the amount of sporulation. Sporangiophores are up to 1500 μm in length and 18 μm in width, smooth-walled, non-septate, simple or branched, arising from stolons, opposite to rhizoids, usually in groups of three or more.

Sporangia are globose, often with a flattened base, grayish-black, powdery in appearance, up to 175 μm in diameter, and many spores are present in them. Columellae and apophysis together are globose, sub-globose, or oval, up to 130 μm in height, collapsing to an umbrella-like form after spore release. Sporangiospores are angular, sub-globose to ellipsoidal, with striations on the surface, and up to 8 μm in length. Fungal growth is absent at 45° C and there is good growth at 40° C.

4.3. Charcoal Rot

4.3.1. Macrophomina

The fungus causes yellowing and senescence of leaves that remain attached to the stems by the petioles, sloughing of cortical tissues from the lower stem and taproot, and the gray appearance of these tissues due to the abundance of microsclerotia that can result in the premature death of the host plant. The infected, rotted portion turns into a charcoal color, and hence the name charcoal rot. It is caused by the fungus *Macrophomina.*

Macrophomina is characterized by hyaline hyphae with thin walls and light brown or dark brown hyphae with septa. Branches from the main hyphae are generally formed at a right angle to the parent hyphae, with constriction at the point of origin. Microsclerotia, a compact mass of hardened fungal mycelium, are spherical, oval, or oblong, light brown in the young stage, becoming darker (brown to black) with age. Pycnidia, which are rarely observed under natural conditions, are larger than microsclerotia, dark brown to black, rough, globose, or irregular, beaked, and ostiolated. The fungus can show a wide heterogeneity in mycelium color, microsclerotia distribution, pycnidia formation, and chlorate phenotypes between isolates on synthetic media. It has been suggested that the morphological heterogeneity could be attributed to the responses of the fungi to environmental factors or variations in their host's species. The important species is *Macrophomina phaseolina.*

4.3.1.1. Macrophomina phaseolina

Colonies are initially whitish on the PDA medium and turn dark brown-grayish in color with age. Hyphal branches generally form at right angles to parent hyphae, but branching is also common at acute angles. Aerial mycelium with completely or partially appressed growth may or may not be formed in culture. Some isolates form concentric growth rings.

Pycnidia are initially immersed in host tissue, then erumpent at maturity. They are 100–200 μm in diameter, dark to grayish, becoming black with age, globose or flattened globose, membranous to sub-carbonaceous, with an inconspicuous or definite truncate ostiole. The pycnidia bear simple, rod-shaped conidiophores, 10–15 μm long. Conidia (14–33 × 6–12 μm) are single-celled, hyaline, and elliptical or oval.

Microsclerotia of *M. phaseolina* are jet black in color and appear smooth and round to oblong or irregular. It varies in size and shape on different substrates. They are formed from aggregates of hyphal cells joined by a melanin material with 50–200 individual cells composing an individual microsclerotia. Semi-appressed mycelium is observed on the culture plate with microsclerotia imbedded within the hyphae or engrossed in the agar. Aggregation of hyphae form microsclerotia with 100–120 µm in size. Microsclerotia are smooth and irregular in shape, and some are round to oblong (Fig. 4.17).

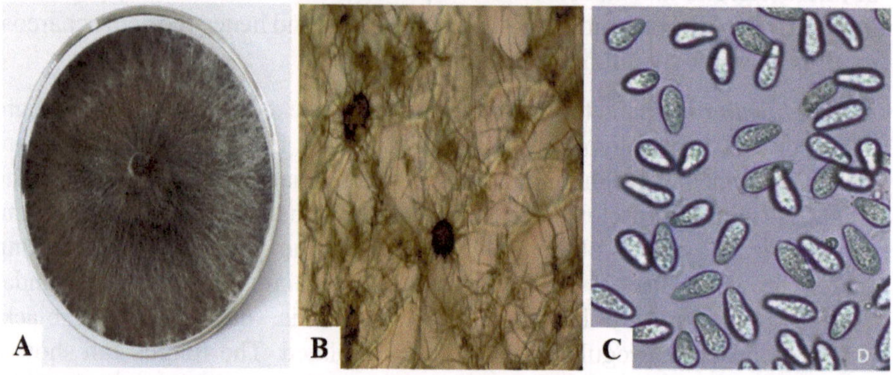

Fig. 4.17: *M. phaseolina fungus*; **(A)** fungal growth on the agar plate, **(B)** microsclerotia, **(C)** conidia.

4.4. Wilt

4.4.1. Fusarium

Fusarium-infected plants are usually stunted; their leaves turn pale green to golden yellow and later wilt, wither, die, and drop off progressively upward from the stem base. Dark streaks occur in the xylem vascular tissue of the roots and lower stem, and the roots may decay. Infected seedlings wilt and die. *Fusarium solani* and *F. oxysporum* are among the most frequently isolated fungal species from the wilt disease symptoms.

Fusarium oxysporum is a cosmopolitan species that causes severe vascular wilts and root rot diseases in various crops.

Fusarium solani infects potatoes, peas, beans, and members of the cucurbit family such as melon, cucumber, and pumpkin to cause *Fusarium* damping-off, corn rot, fruit rot, root rot, and surface rot.

Fusarium colonies are usually pale or brightly colored (depending on the species) and may have a cottony aerial mycelium. Their color varies from whitish to yellow, brownish, pink, or reddish. Species of *Fusarium* typically produce spores (called macro- and micro-conidia) for reproduction and dissemination.

The identification of *Fusarium* species is mainly based on distinctive characteristics of the shapes and sizes of macro- and micro-conidia, the presence and absence of chlamydospores, as well as colony appearances, pigmentations, and growth rates on agar media (Figs. 4.18 and 4.19).

The difference in *F. oxysporum* and *F. solani* in different parameters is given in Table 4.1.

Table 4.1: Characteristics of the microscopic structures of *Fusarium solani* and *F. oxysporum*.

Parameters	***Fusarium solani***	***Fusarium oxysporum***
Color of the colony (on PDA)	Light yellow, moist appearance, red with cottony and orange brown mycelium, with light brown exudate	Light and dark violet, salmon- colored, purplish brown, all with cottony mycelium, without exudate
Growth (on PDA)	4–5 days	4–5 days
Type of micro-conidia (on CA)	(8–16 × 2–4 μm)	Extended ovals (5–12 × 2.2–3.5 μm)
Type of macro-conidia (on CA)	Abundant, cylindrical, dorsal and ventral surface parallel, 3–5 septa (35–55 × 4.5–6 μm)	Abundant, sickle-shaped, 3–5 septa with distinctive cellular foot (27–55 × 3–5 μm)
Types of chlamydospore (on CA)	Single or in pairs	Abundant, single or in pairs
Type of phialide (on CA)	Long and thin, approx. 100 μm or more	Short robust mono-phialides in straight or vertical angles <20 μm

PDA: potato dextrose agar (Tuite 1969); CA: carnation agar (sterile carnation leaves on water agar.

Different morphotypes are also known in these two wilt pathogens which differ in different parameters (Table 4.2).

Table 4.2: Morphological characteristics of morphotypes I and II of *(A). Fusarium solani* isolates, and (B*). Fusarium oxysporium* isolates.

***A. F. solani* morphotypes**		
	Morphotype I	Morphotype II
Macroscopic characteristic		
Colony color[a]	Pale to brown, brown-greenish to white-greenish	White-creamy to white-grayish
Pigmentation[b]	Pale brown, yellowish brown with dark brown zonation	Colorless, white-creamy with dark brown zonation
Growth rate (cm)[c]	3.3 ± 0.2	3.5 ± 0.2
Microscopic characteristic		
Mean length of macroconidia (μm)[d]	34.4 ± 9.8	42.0 ± 3.0
Mean width of macroconidia (μm)[d]	3.7 ± 0.5	4.7 ± 0.5
Conidiogenous cell	Long and branched monophialides	Long monophialides
Production of chlamydospores	Sparsely produced on SA	Abundantly produced on CLA
Presence of sporodochia	Cream and blue	Cream
Macroconidia septation	3–5	3–7
***B. F. oxysporium* morphotypes**		
	Morphotype I	Morphotype II
Macroscopic characteristic		
Colony color[a]	Pale, dark to peach-violet	White to white-violet
Pigmentation[b]	Pale, dark to peach-violet	Pale violet to colorless
Growth rate (cm)[c]	4.3 ± 0.7	4.5 ± 0.5
Microscopic characteristic		
Macroconidia morphology	Straight and relatively slender	Slightly curved and thick
Apical cell morphology	Tapered and curved	Tapered with slight hook
Basal morphology	Pointed	Foot shaped
Mean length of macroconidia (μm)[d]	42.2 ± 4.5	34.9 ± 2.1

***B. F. oxysporium* morphotypes**		
	Morphotype I	Morphotype II
Mean width of macroconidia (μm)[d]	3.9 ± 0.4	4.2 ± 0.3
Macroconidia septation	3–7 (three-septate most common)	3–5 (three-septate most common)

[a]Colony color were determined by observing the upper surface of the colony.
[b]Pigmentation were determined by observing the lower surface of the colony.
[c]Growth rates were taken after three days of incubation at 25° C.
[d]Mean values of length and width of 50 randomly picked macroconidia ± standard deviation.

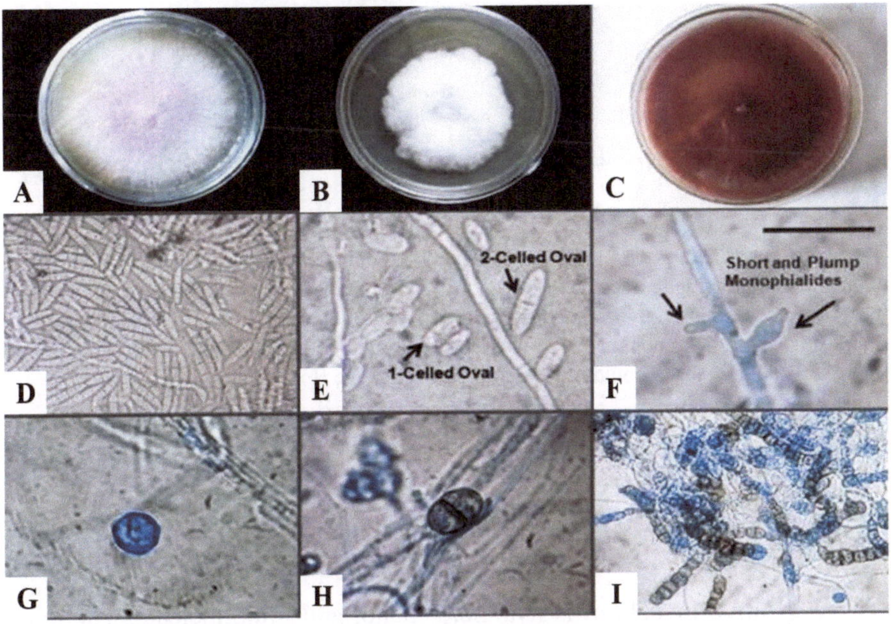

Fig. 4.18: *Fusarium oxysporum.* **(A–C)** colony morphology on potato dextrose agar medium, **A:** pinkish colony color, **B:** fluffy white growth, **(C)** dark violet pigmentation on the under surface of plate; **(D)** macroconidia; E: microconidia of single- and two-celled oval shape; **(F)** conidiogenous cells; **(G–I)** formation of chlamydospores, **(G)** singly, **(H)** pairs, **(I)** short chains, **(D–I)** *scale bar* = 25 μm.

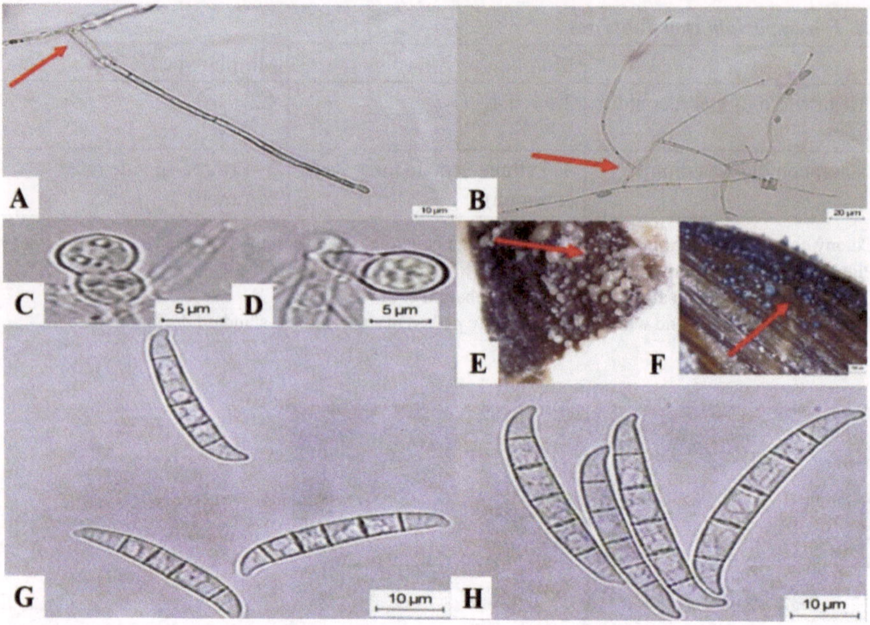

Fig. 4.19: *Fusarium solani.* **(A)** long monophialides commonly observed in morphotype II; **(B)** branched long monophialides observed only in morphotype I, **(C, D)** typical chlamydospores observed in both morphotypes I and II; **(E)** cream sporodochia produced by both morphotypes I and II; **(F)** blue sporodochia observed only in morphotype II; **(G)** macroconidia produced by morphotype I; **(H)** macroconidia produced by morphotype II.

Source: R. Hafizi, B. Salleh, and Z. Latiffah, School of Biological Sciences, Universiti Sains Malaysia, Penang, Malaysia.

4.4.2. Verticillium

The vegetative mycelium is hyaline, septate, and multinucleate. The nuclei are haploid. Conidia are ovoid or ellipsoid and usually single-celled. They are borne on phialides, which are specialized hyphae produced in a whorl around each conidiophore. Each phialide carries a mass of conidia. *Verticillium* is named for this "verticillate" (=whorled) arrangement of the phialides on the conidiophore.

As the diseased plant senesces, the fungus becomes saprophytic and colonizes the dying tissues. During colonization, the fungus forms microsclerotia, which are masses of melanized hyphae (Fig. 4.20).

Fig. 4.20: Fungal culture of *Verticillium*, spores borne on phialide and cross section of infected stem with visible internal vascular circular tissue discoloration.

Verticillium dahliae, *V. albo-atrum*, *V. longisporum*, *V. nubilum*, *V. theobromae*, and *V. tricorpus* are the different species belonging to the *Verticillium* genus that cause wilt disease. *V. albo-atrum* only forms mycelium, while *V. dahliae* and *V. longisporum* produce microsclerotia, *V. nubilum* forms chlamydospores, and *V. tricorpus* forms mycelium, microsclerotia, and chlamydospores.

4.4.3. *Ceratocystis*

The *Ceratocystis* species causes vascular wilt of woody plants like oak, cocoa, and eucalyptus and some woody fruit trees like pomegranate. The fungus produces sporulation mats emanating fruity odors and forming asexual barrel-shaped spores in chains called endoconidia. Especially, certain species of sap beetles, birds, and other animals are attracted to the fruity smell. On these mats, perithecia with long beak are formed producing sexual ascospores. Conidia and ascospores are mainly dispersed by air, rain, or insects such as the sap-feeding vector *Colopterus truncates*. The important species is *Ceratocystis fimbriata*.

4.4.3.1. Ceratocystis fimbriata

Morphological characterization of *Ceratocystis fimbriata* is based on production of aleurioconidia, endoconidia, ascospore, and perithecia. On oat meal agar, the colony color of *C. fimbriata* is grayish with flat type of colony growth and regular type of margin. The margin color is grayish and colony diameter is 90 mm after 16 days of incubation at room temperature. The color of colony changes to grayish color with age owing to production of aleurioconidia, endoconidia, ascospores, and perithecium.

Black colored perithecia with globose base is observed with size of 181.1 × 131.2 μm, exuding small, hyaline and hat shaped ascospores from the apex of the perithecium which measure 5.13 × 4.27 μm. Endoconidia are hyaline, cylindrical, and average size is 23.6 × 4.90 μm. Aleurioconidia are thick walled ellipsoidal or pyriform with size of 18.5 × 10.10 μm (Fig. 4.21).

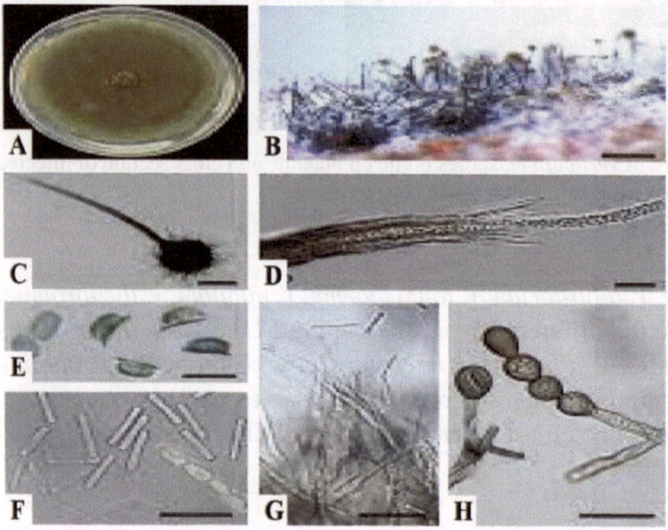

Fig. 4.21: *Ceratocystis.* **(A)** mycelial growth in petriplate; **(B)** mycelial mat with fruity order at infected site, **(C)** perithecia with long beak; **(D)** ostiole with ascospore release; **(E)** ascospores; **(F–H)** endoconidia and aleuriconidia.

The optimum temperature for the maximum mycelial growth of *C. fimbriata* is 22.5° C. The optimal temperature for the maximum sporulation is 22.4° C. There is no mycelial growth below 8° C and above 36° C. The perithecium is superficial or submerged in the substrate, with color ranging from brown to black both at the base and neck of the perithecium. The ostiolar hyphae is divergent, light brown to hyaline, aseptate. The ascospores are hyaline, small, hat-shaped and accumulates in a cream-colored matrix at the end of the perithecium. *C. fimbriata* has the highest values for mycelial growth in the PDA and V8-agar media.

4.4.4. Cephalosporium

Late wilt, or black bundle disease, is a vascular wilt disease of maize caused by the soil-borne and seed-borne fungus, *Harpophora maydis*. Synonyms are *Cephalosporium maydis* and *Acremonium maydis*. The disease is characterized by the presence of black bundle of hyphal rope in the infected portions.

On CMA and PDA solid media, the mycelial colonies reach 90 mm diameter within eight days at 28° C under continuous cool-white fluorescent lighting. Mycelial colony growth is faster in the dark than in light conditions, covering a 90 mm diameter plate within six days.

Mycelial colonies are flat, white at first, then turning grey to black with age. The margins of the older (10–21 days) colonies have a "rhizoidal" or "hyphal rope" appearance and they have typical sclerotium-like bodies, aggregates of thick-walled, dark cells.

Hyphae are 2–4.5 µm in width, hyaline, septate, branched, and decumbent. Sporulation occurs within a few days of transferring colonies to new plates. Conidiophores are also hyaline, septate and branched, and each carried four to seven single-celled, oval to cylindrical, hyaline, smooth-walled conidia. Conidia are 8.5–25 µm long (average 15 µm) and 5–7 µm wide. Conidia germinates rapidly, usually bipolar germ tubes are developed, but sometimes one or three germ tubes are formed. Anastomosis of germ tubes often occurs (Fig. 4.22).

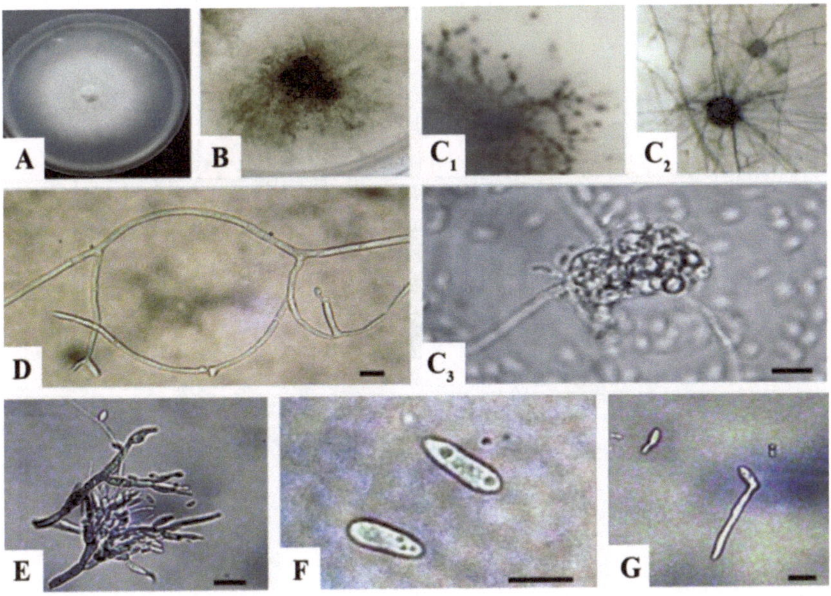

Fig. 4.22: *Cephalosporium maydis* strain Hm-3. **(A)** young mycelial colony (six days old on PDA); **(B)** old mycelial colony (12 days old) exhibiting a "rhizoidal" or "hyphal rope" appearance; **(C)** sclerotium-like bodies formed in old cultures (44 days old) on PDA (seen as spots from the back of the Petri dish C_1. Magnified on the culture surface C_2 and at high magnification in a slide preparation C_3); **(D)** bent marginal hyphae; **(E)** two conidiophores; **(F)** conidia; **(G)** germinating conidia after 6 h of incubation.

4.5. Wart

4.5.1. Synchytrium endobioticum

Synchytrium endobioticum is an obligate parasite that infects several plants of the genus *Solanum. S. endobioticum* develops no mycelium. The fungus produces a thick-walled structure known as a winter sporangium. It is 25–75 µm in diameter and contains 200–300 spores. Sporangia are clustered into thin-walled sori. The motile life stage of the zoospore is about 0.5 µm in diameter and has one posterior flagellum. Thin-walled, transparent summer sporangia are formed in young warts during the growing season. Resting spores are found during the growing season and located in decomposing warts; they are embedded in the host tissue, filling the host cell almost completely.

Resting spores are aseptate, golden brown and thick-walled (triple wall) with the outer wall furrowed, prominently ridged, and irregularly thickened, 25–75 µm (mean 50 µm) in diameter. They are spherical to ovoid in shape. When warts decay, the resting spores initially carry some attached host cell tissue, which eventually disintegrate and the resting spores are left with a characteristic angular appearance (Fig. 4.23).

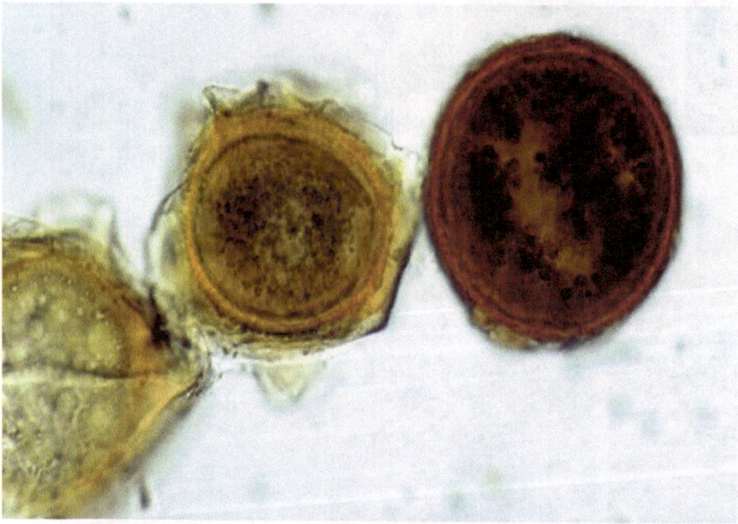

Fig. 4.23: Resting spores of *Synchytrium endobioticum* (left) and *Synchytrium succisae* (right). *Source*: Photo courtesy by SASA, GB.

4.6. Club Root

4.6.1. Plasmodiophora brassicae

The infection of *Plasmodiophora brassicae* leads to excessive proliferation of infected host cells in the form of tuberous swellings or malformations on the diseased roots. These malformations are called clubs or galls. This tumorous growth (galls or clubs) of roots and hypocotyl resulting from cell hypertrophy and hyperplasia is the most striking feature of *Brassica* plants infected with *P. brassicae*.

During the life cycle of the pathogen two distinct generations or phases *viz.* gametophyte generation (or haplophase) and sporophyte generation (or diplophase) are observed. The former occurs in the root hairs and the latter in the cortical cells of the host root (*Brassica* sp.).

The gametophyte generation (haplophase) is initiated by the differentiation of resting spores (sporangia) or cysts by meiosis from the diploid protoplast of the secondary plasmodium in the club roots of the host cortex. The mature cysts, resting sporangia, or spores which are haploid, are released into the soil when the club roots undergo decay and decomposition by soft-rot microorganisms. After liberation, they perennate or germinate immediately if the conditions for growth are favorable. On germination, each resting spore or cyst produces a biflagellate, heterokont, or haploid zoospore. The released zoospore swims in a thin film of water in the soil or creeps over the soil particles like an ameba. When it comes into contact with the root hairs of a suitable host it becomes attached and ameboid. In this state, it is called a myxamoeba. It then retracts its flagella and enters the root hair. Within the root hair, it is swept away from the penetration site by host cytoplasmic cyclosis. Through growth and repeated cruciform nuclear division, the uninucleate myxamoeba becomes converted into a naked, multinucleate thallus-like mass of protoplasm. It is an acellular haploid and is called the primary plasmodium (or gametothallus).

Reaching a certain stage of maturity, the haploid protoplast of the gamete thallus cleaves into 4–10 small, spherical, thin-walled, uninucleate gametia. The haploid nucleus of each gametangium undergoes repeated mitosis to form 4–8 nuclei. Each daughter nucleus gathers a bit of cytoplasm around it to become a spindle-shaped motile biflagellate isogamete. All the above-mentioned haploid stages, *viz.*, the resting meiospores (sporangia or cysts), the zoospores, the primary plasmodium (gametothallus), the gametangia, and the isogametes (which occur in the root hairs of the host), constitute the gametophyte generation, or haplophase.

The second phase of the parasite life cycle, commonly called the diplophase or sporophyte generation, starts with the formation of zygotes. The zygotes are found in the cortical cells of the root. It is not definitely known whether the isogametes fuze in the root hairs or in the cortical cells after migration. The consensus favors the view that the isogametes are released in the soil outside the host, where pairing followed by plasmogamy takes place. Karyogamy is delayed. A binucleated quadriflagellate zygote thus formed reinfects the host root. Within the root cortex, the zygote retracts its flagella and becomes ameboid in form. It is called myxamoeba. Within the invaded cell, the myxamoeba, by subsequent growth accompanied by repeated cruciform division, grows to form a naked, multinucleate diploid mass of protoplasm called the secondary plasmodium, or sporothallus. It fills the lumen of the cortical cell. It then ceases to grow in size, and its diploid protoplast undergoes meiosis. As a result, the multinucleate diploid protoplast of the secondary *plasmodium* is converted into a large number of uninucleate spherical haploid bodies, each of which is surrounded by a wall of its own. These are the resting spores (sporangia) or cysts.

The resting spores, when mature, are set free in the soil by the decay of the club by rot microorganisms. In the soil, the resting spores or cysts germinate under favorable conditions to continue the fresh life cycle. The zygote and the secondary plasmodium constitute the diplophase, or sporophytic generation.

Alternation of generations of Plasmodiophora brassicae

There is distinct alternation of generations in the life cycle of *Plasmodiophora*. The zygote and the secondary plasmodium (sporothallus before spore formation) represent the sporophyte generation or diplophase. The resting spores, the zoospores, the haploid primary plasmodium (gametothallus), the gametangia, and the isogametes represent gametophyte generation or haplophase (Fig. 4.24).

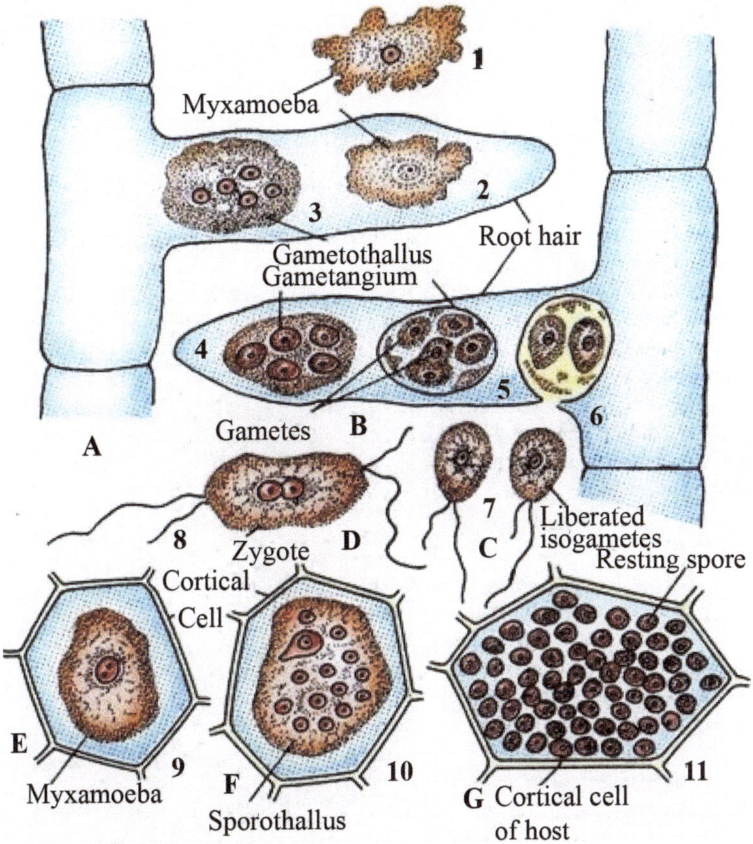

Fig. 4.24: *Plasmodium brassicae.* **(A)** 1. myxamoeba outside the root hair, 2. within the root hair, 3. gametothallus gametangium. **(B)** 4–6. differentiation of gametes in the gametangium; **(C)** liberated isogametes, **(D)** fuzing gamete and formation of zygote; **(E)** zygote or myxamoeba in the host cortical cell, **(F)** sporothallus, **(G)** resting sporangia or spores in the host cell.

The two crucial points in the life cycle are meiosis at the time of differentiation of spores (meiospores) and karyogamy (fusion of gametic nuclei). At these critical points, the life cycle switches from one generation to the next. With meiosis, the diploid or sporophytic phase ends. The gametophyte phase starts. At the time of karyogamy, the gametophyte phase terminates and the sporophytic phase is initiated with the formation of diploid zygote. These two generations occur regularly, one after the other, in the life cycle. This is called the alternation of generations. The life cycle characterized by the alternation of generations with sporogenic meiosis is called haploid-diploid or diplohaplontic (diplobiontic life cycle).

5

Identification of Fungi Causing Diseases Near Collar Region

5.1. Gummosis

5.1.1. Phytophthora

Phytophtora citrophthora, *Phytophtora nicotianae*, *Phytophtora syringae* are the *phytophthora* species associated with citrus gummosis; while *Phytophtora pistaciae sp. nov.* and *Phytophtora melonis* are the principal causes of pistachio gummosis.

5.1.1.1. Phytophthora citrophthora

Phytophtora citrophthora is characterized by a finely radiating, white rosette and slightly cottony colonies. Hyphae are 6–7 µm in diameter. Sporangia are mostly laterally attached and are about 40 × 30 µm. Shapes ranged from spherical, ovoid, obpyriform, and obturbinate, to ellipsoidal, but mostly papillate. Some are asymmetrical, and others often have two divergent apices. Papillae measure up to 4 µm long with an average length-to-breadth ratio of 1.4:1. Sporangiophores are irregularly branched singly or in a loose sympodium with a swelling at the point of branching.

Colonies and cardinal temperatures

Phytophthora citrophthora will grow in the chrysanthemum pattern within seven days when grown on V8 agar, potato dextrose agar (PDA), and malt extract agar (MEA) media. Minimum growth temperature is 6° C, optimum temperature range is 24–28° C, and maximum temperature is 33° C.

Asexual phase

Sporangia are semi-papillate, some with two papillae, persistent, ovoid, limoniform, ellipsoid, globose, and with distorted shapes (17–48 × 20–110 µm), some with tapered bases, and with intercalary position, borne in simple sympodial or irregularly branched sporangiophores. Chlamydospores are globose, sub-globose, lateral, terminal, and intercalary (24–37 µm diameters). Hyphal swellings are absent.

Sexual phase

Sterile-heterothallic fungi. Oospores are rarely produced.

Most typical characteristics

Phytophthora citrophthora is characterized by the shapes of the semi-papillate, persistent sporangia and the chlamydospores (Figs. 5.1–5.3).

5.1.1.2. Phytophthora nicotianae

Colonies and cardinal temperatures

The colony morphology of *P. nicotianae* on V-8 agar and potato dextrose agar has no distinct pattern and has coralloid edges. On malt extract agar, this fungus shows slow growth. The minimum temperature for growth is 9° C, the optimum temperature is 24° C, and the maximum temperature is 33° C.

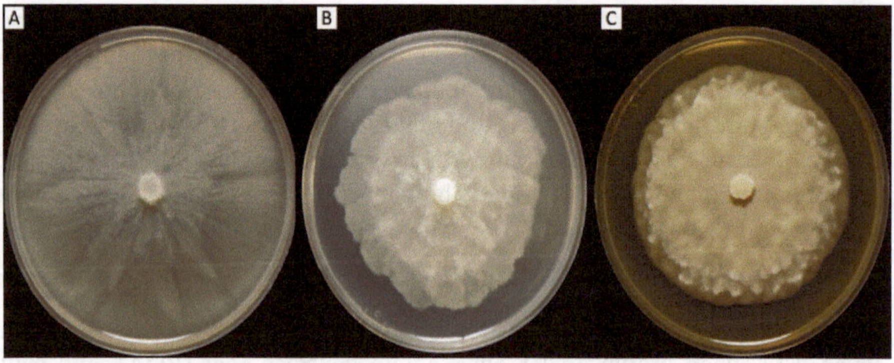

Fig. 5.1: *Phytophthora citrophthora.* Colony morphology at 7th day on **(A)** V-8 agar, **(B)** potato dextrose agar, and **(C)** malt extract agar media.
Source: Photo courtesy by Krysta Jennings and Leandra Knight, USDA-APHIS-PPQ.

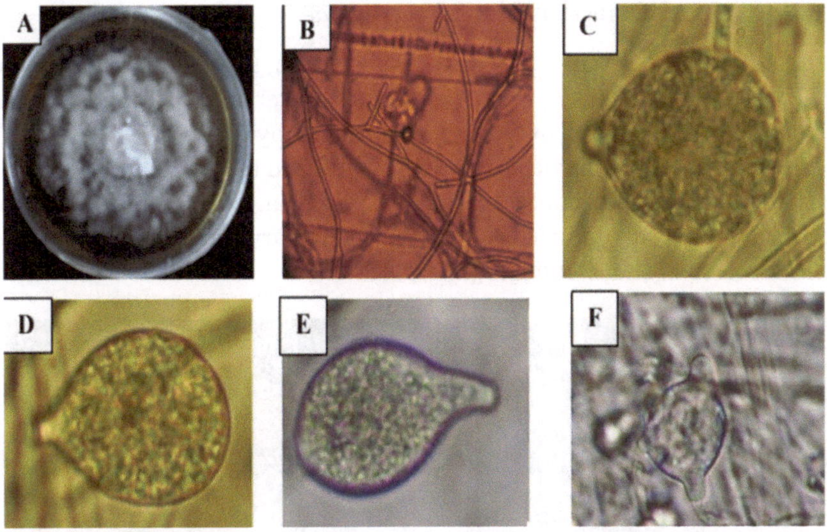

Fig. 5.2: Morphological characteristics of *Phytophthora citrophthora*. **(A)**: finely radiate cottony mycelium, **(B)**: coenocytic hyphae of up to 6–7 nm in diameter, **(C and D)**: ovoid and asymmetric sporangium, **(E)** sporangium with prominent papilla. **(F)**: sporangium with more than one papilla. *Bar* = 10 μm.

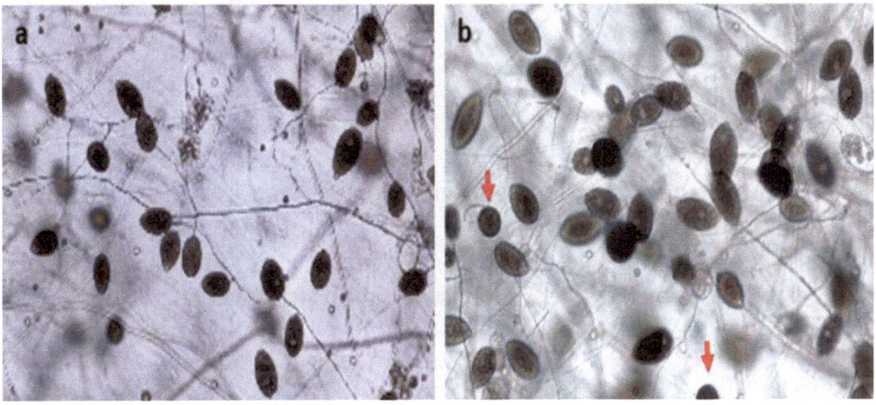

Fig. 5.3: Asexual phases (sporangia and chlamydospore) formed on V8 agar flooded with soil extract. **(A)** semi-papillate persistent sporangia originated in irregular branched and simple sympodial sporangiophores; **(B)** semi-papillate persistent sporangia and globose chlamydospores (marked with red arrows).
Source: Photo courtesy of Gloria Abad, USDA-APHIS-PPQ.

The morphological characteristics of *P. nicotianae*

Mycelium is a dense or loose rosette with no pattern, spreading, or arachnoid aerial. Hyphae are coenocytic, with an average diameter of 7–10 μm. Sporangial shapes range from ellipsoid, ovoid, pyriform, and obpyriform to spherical with a prominent papilla. Sporangia size is 36 × 28 μm and the length–breadth ratio is 1.34:1. Some isolates produce intercalary sporangia. Chlamydospores, 13 to 60 μm in diameter, are produced abundantly intercalary and terminally, and no sexual structures in single cultures (Fig. 5.4).

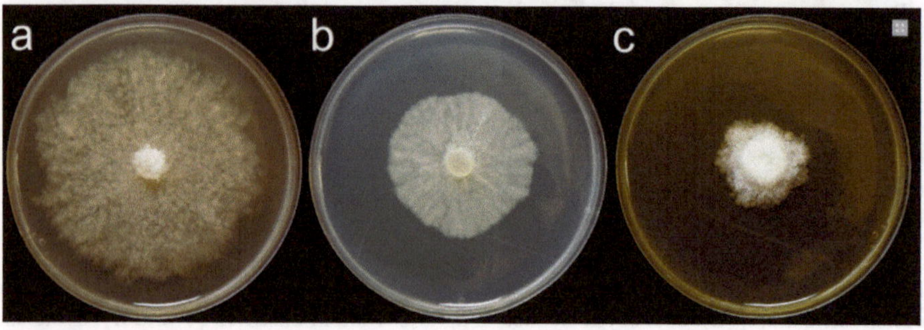

Fig. 5.4: *Phytophthora nicotianae* colonies grown for seven days on **(A)** V8 Agar, **(B)** potato dextrose agar, and **(C)** malt extract agar media.
Source: Photo courtesy by Krysta Jennings and Leandra Knight, USDA-APHIS-PPQ.

Asexual phase

Sporangia are papillate; persistent and caducous with short pedicels (1-4 μm); ovoid, globose, obovoid, obpyriform, and with irregular shapes (15–64 L × 11–47 W μm); some sporangia are with elongated necks and tapered bases; sporangia are originated in simple sympodial sporangiophores. Hyphal swellings are globose, sub-globose, intercalary, and often have radiating hyphae. Chlamydospores are globose (15–43 μm), lateral, terminal, and intercalary.

Sexual phase

Heterothallic fungi. Oogonia is smooth-walled (24–32 μm diameters); antheridia are amphyginous (9–21 × 10–15 μm); and oospores **are** predominantly aplerotic or slightly aplerotic (19–27 μm diameters).

Most typical characteristics

Phytophthora nicotianae is characterized by the presence of asexual phases, including sporangia, hyphal swellings, chlamydospores, and the typical coralloid hyphae observed in the culture media (Fig. 5.5).

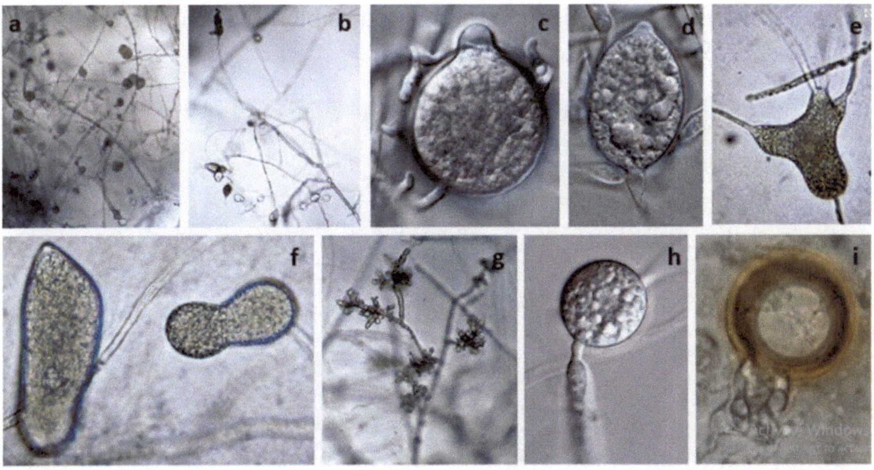

Fig. 5.5: *Phytophthora nicotianae* asexual phase **(A–H)** and sexual phase (i): **(A, B)** sporangia originated in simple sympodial sporangiophore; **(C, D)** papillate sporangia; **(D)** caducous with short pedicel; **(E, F)** sporangia with irregular shapes; **(G)** hyphal swellings; **(H)** chlamydospore; **(I)** oogonium with amphigynous antheridium and aplerotic oospore.
Source: Photo courtesy by Gloria Abad, USDA-APHIS-PPQ.

5.1.1.3. Phytophthora syringae

Colonies and cardinal temperatures

Colony on CMA (cornmeal agar) media has no special pattern while on PDA (potato dextrose agar) and V-8 agar media it has chrysanthemum and rosette pattern. The minimum temperature for growth is 5° C, the optimum temperature range is 18–20° C, and the maximum temperature is 23–25° C. Oogonia are formed abundantly in culture and host tissues.

Asexual phase

Sporangia are semi-papillate, persistent, broad ovoid, ellipsoid, obpyriform, or distorted shapes (20–65 μm length × 18–40 μm width) sometimes with tapered base; they originated in close simple sympodial sporangiophores with intercalary swellings. Hyphal swellings are often catenulated and spherical or irregular in shape. Chlamydospores are absent.

Sexual phase

Homothallic fungi. Oogonium are smooth-walled, globose (25–47 μm diameters), sometimes with tapered bases. Antheridia are spherical, ovoid, or club-shaped and are predominately paragynous and rarely amphigynous, sometimes with digitate projections and oospores plerotic, aplerotic, or slightly aplerotic (20–39 µm diameters) (Fig. 5.6).

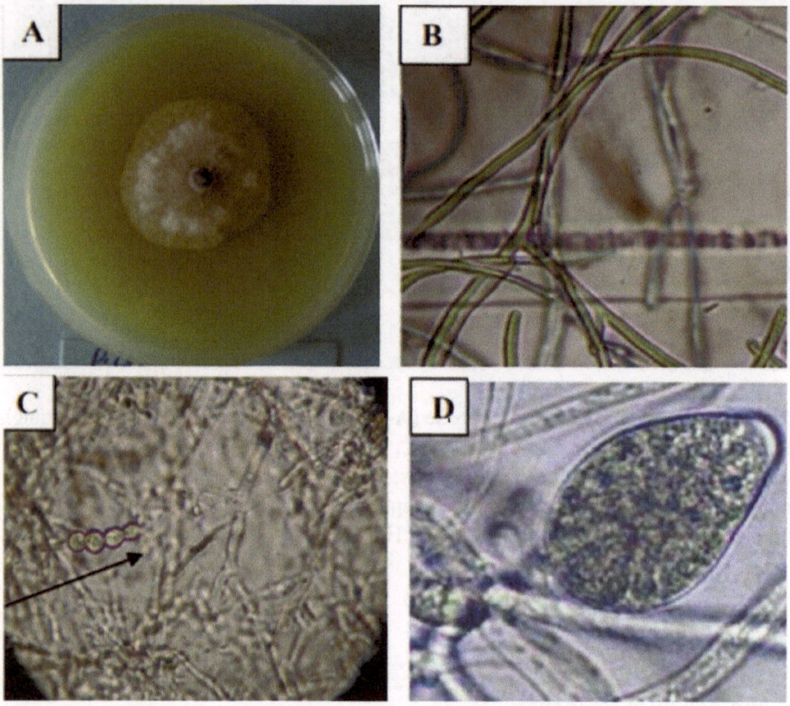

Fig. 5.6: *Phytophthora syringae*. **(A)** dense or loose rosette spreading aerial mycelium with no pattern; **(B)** slender coenocytic hyphae; **(C)** with rounded or angular hyphal swellings, often in chains (catenulate) and delimited by septa; **(D)** Sporangia are broadly ovoid or obpyriform, semi-papillate, persistent, and formed in succession from a single sporangium. Average sporangial dimension is 39 × 26.5 µm, with the length–breadth ratio varying from 1.32:1 to 1.85:1. No chlamydospores are produced on agar media.

Differentiating characteristic between P. citrophthora, P. nicotianae vs P. syringae

The species *P. citrophthora* and *P. nicotianae* exhibit papillated sporangia, which is the main characteristic that differentiates them from *P. syringae*, which has semi-papillated sporangia. The presence of rounded or angular hyphal swellings that are often in chains (catenulate) and delimited by septa is unique to *P. syringae* and facilitates its differentiation from *P. nicotianae* and *P. citrophthora.* The latter two species, i.e. *P. citrophthora* and *P. nicotianae*, are heterothallic and thus did not produce oospores or sexual structures in single cultures.

The optimal temperature for growth is another main distinguishing feature between them. The optimum temperatures of *P. citrophthora* are 24–28 °C and those of *P. syringae* and *P. nicotianae* are 20 and 30 °C, respectively.

5.1.1.4. Phytophthora pistaciae

Colonies and cardinal temperatures

Colony morphology on PDA, V-8, and MEA media has no distinct pattern. Minimum growth temperature for this fungus is 9° C, the optimum temperature is 27° C, and the maximum temperature 33° C.

Asexual phase

Sporangia are non-papillate, persistent, ellipsoid, ovoid, obpyriform, and pyriform (41–75 L × 18–34 W μm), mostly with tapered bases, showing internal proliferation (nested), and originated in unbranched sporangiophores. Hyphal swellings are globose or sub-globose with radiating hyphae. Chlamydospores are absent.

Sexual phase

Homothallic fungi. Oogonia are smooth-walled (25–44 μm diameters), antheridia are predominantly paragynous, spherical, ovoid, or ellipsoid, showing knotted hyphae or digital projections, and oospores are aplerotic (21–33 μm diameters).

Most typical characteristics

Phytophthora pistaciae is characterized by the presence of homothallic oogonia with antheridia radiating knotted hyphae and the presence of globose or sub-globose hyphal swellings with radiating hyphae (Fig. 5.7).

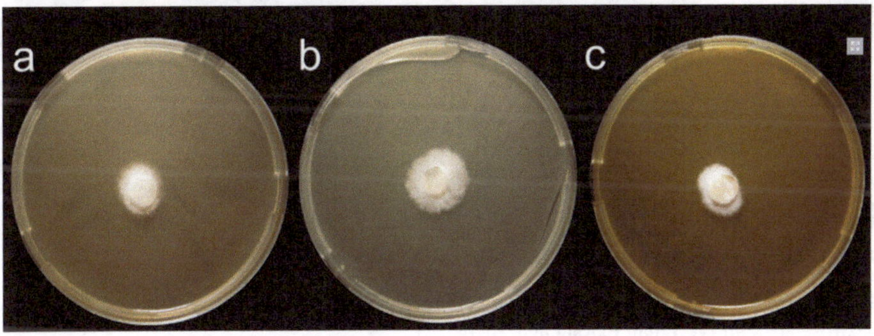

Fig. 5.7: *Phytophthora pistaciae.* Mycelial colony at seven days on **(A)** V-8 agar, **(B)** potato dextrose agar, and **(C)** malt extract agar media.
Source: Photo courtesy by Krysta Jennings and Leandra Knight, USDA-APHIS-PPQ.

5.1.1.5. Phytophthora melonis

Colonies and cardinal temperatures

Colony morphology after seven days of growth on V-8 agar, potato dextrose agar, and malt extract media with no distinct pattern to slightly radiate (fig 5.8). Minimum temperature for growth of the fungus is 9° C, optimum temperature range 28–32° C, and maximum temperature 37° C.

Asexual phase

Sporangia are non-papillate with internal proliferation, persistent, sometimes caducous with medium to long pedicels, ovoid, obpyriform with round bases (32–75 × 23–45 μm diameters), produced in long, unbranched sporangiophores. Hyphal swellings are occasionally produced, globose, terminal, and intercalary, some with radiating hyphae. Chlamydospores are spherical (16–52 μm diameters).

Sexual phase

Heterothallic fungi. Oogonia are smooth-walled (27–35 μm diameters). Antheridia are amphigynous (10–29 × 10–20 μm), and oospores are plerotic to slightly aplerotic. The "redescription of *Phytophthora melonis*" indicates that the species doesn't produce chlamydospores or non-papillate sporangia.

Most typical characteristics

Phytophthora melonis is characterized by the presence of non-papillate sporangia with internal proliferation (Fig 5.9).

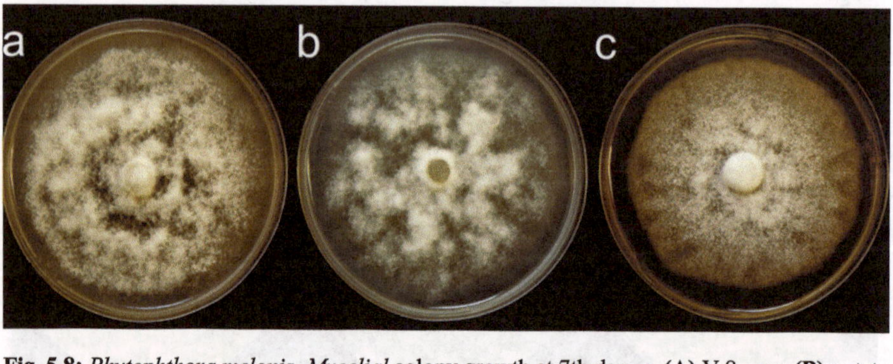

Fig. 5.8: *Phytophthora melonis. Mycelial* colony growth at 7th day on **(A)** V-8 agar, **(B)** potato dextrose agar, and **(C)** malt extract agar media.

Source: Photo courtesy by Krysta Jennings and Leandra Knight, USDA-APHIS-PPQ (Fig. 5.9).

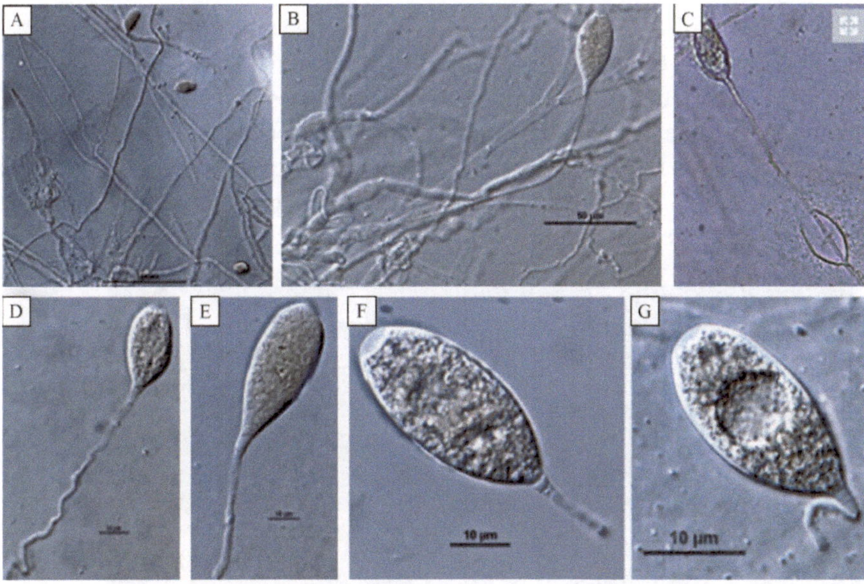

Fig. 5.9: Asexual phase shows non-papillate sporangia **(A–G)**, sometimes caducous with medium to long pedicels **(E–G)**, hyphal swellings **(A, B)**, and internal proliferation **(C)**.
Source: Photo courtesy by Gloria Abad USDA-APHIS-PPQ.

5.1.1.6. Phytophthora sinensis

Its growth is sparse and submerged in all media; growth ceases at 35° C. Chlamydospores are present; oogonial stalks are often funnel-shaped; oogonia are often aborted; and is small in size.

5.1.1.7. Phytophthora sojae

Colonies and cardinal temperatures

The colony after seven days of growth on potato dextrose agar and malt extract agar media is very small, and on V-8 agar, it grows well with no defined pattern (fig. 5.10). The minimum growth temperature for this fungus is 10° C, the optimum temperature range is 22–30° C, and the maximum temperature is 33° C.

Asexual phase

Sporangia are non-papillate, ovoid, ellipsoid, obpyriform, or obovoid (43–101 L × 24–55 W μm) with a tapered base, originated in simple sympodial and undifferentiated sporangiophores, and sometimes exhibit internal and external proliferation. Hyphal swellings are intercalary, spherical, or irregular in shape. Chlamydospores are absent.

Sexual phase

Homothallic fungi. Oogonia are spherical to sub-spherical (24–45 × 23–41 µm), some with distorted shapes, some with tapered bases and some with tangled or knotted hyphae around the antheridium. Antheridia are predominantly paragynous (9–18 L × 8–16 W µm), frequently with a single long spine or digitate projection. Oospores are plerotic, aplerotic, and slightly aplerotic (18–39 µm diameters) (fig 5.11).

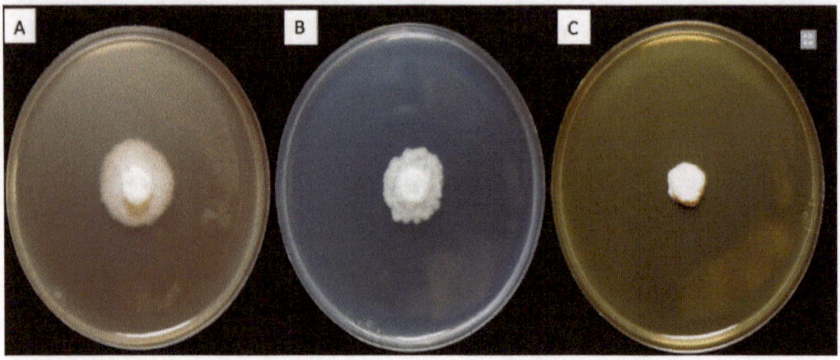

Fig. 5.10: *Phytophthora sojae.* Mycelial colony growth at 7th day on **(A)** V-8 agar, **(B)** potato dextrose agar, and **(C)** malt extract agar media.
Source: Photo courtesy by Krysta Jennings and Leandra Knight, USDA-APHIS-PPQ.

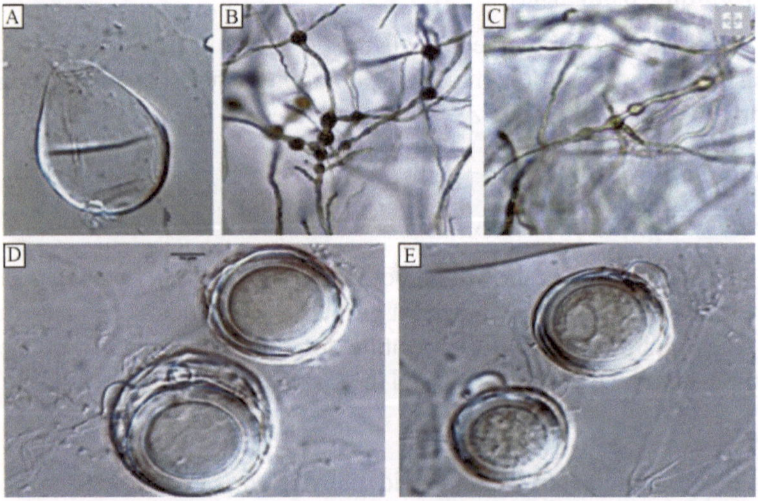

Fig. 5.11: Asexual **(A–C)** and sexual **(D, E)** phases of *P. sojae*: **(A)** sporangium, **(B, C)** hyphal swellings, **(D, E)** oogonia with paragynous antheridia and plerotic, slightly aplerotic, and plerotic oopores.
Source: Photo courtesy by G. Abad, USDA-APHIS-PPQ.

5.1.2. Botryosphaeria

Botryosphaeria causes gummosis disease in Japanese apricot and peach trees. A fungal colony develops villous mycelium. The front side of colonies on PDA is initially white and then turns to gray and dark. The transparent mycelium has transverse septa and a smooth wall. Based on the colony and conidial morphology of *Botryosphaeria*, there are different fungal groups and species.

5.1.2.1. Botryosphaeria dothidea

Culture characteristics

Colonies are olivaceous and then turn gray with the reverse side black. Mycelial mat is moderately dense with a smooth margin. The Optimal temperature for fungal growth is 25–30° C. Its colony on the PDA reached a 50-mm radius after four days at 25° C in the dark. Conidia (17–22.6 × 4.3–6.0 μm) are hyaline, unicellular, and spindle-shaped. Asci (78–125 × 15–17 μm) are hyaline, bitunicate, clavate, and eight-spored. Ascospores (18–22 × 7.0–8.2 μm) are hyaline and spindle-shaped or fusoid (Fig. 5.12 & 5.13).

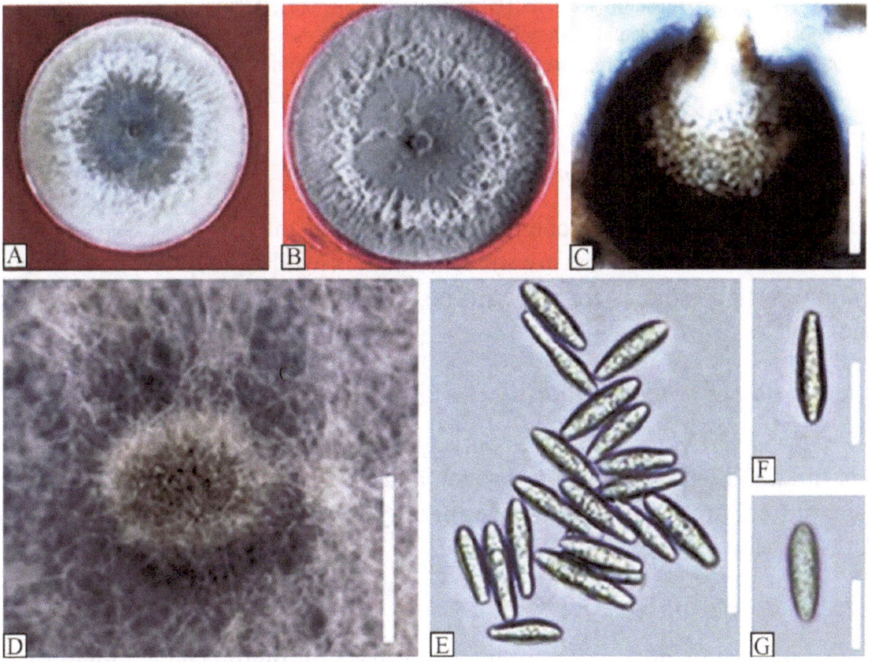

Fig. 5.12: *Botryosphaeria dothidea.* **(A)** young colony on PDA (seven days), **(B)** aged colony on PDA (14 days), **(C, D)** conidiomata on PDA culture. **(E–G)** hyaline conidia. *Scale bars* (C) = 30 μm, (D) = 100 μm, (E–G) = 10 μm.

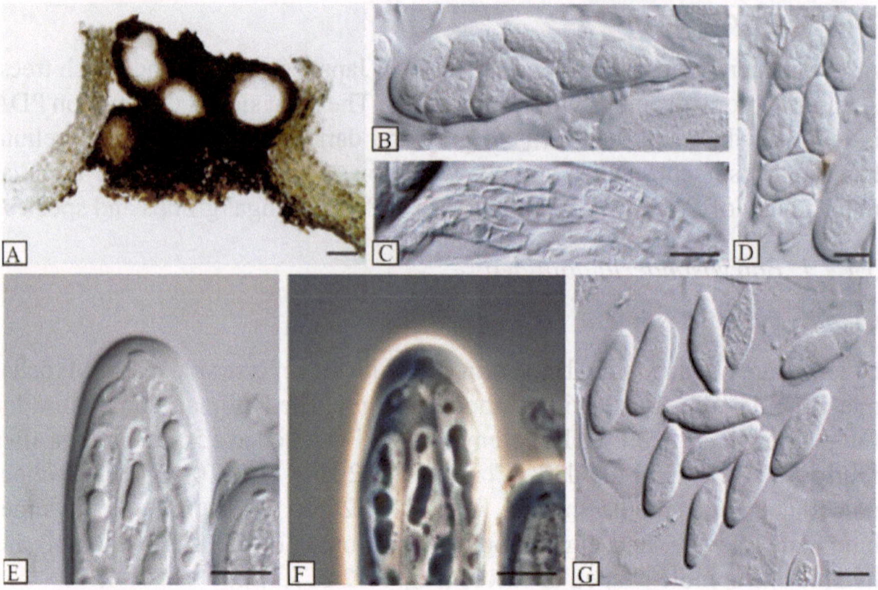

Fig. 5.13: (A) section through ascomata, **(B)** ascus with ascospores, **(C)** septate pseudoparaphyses. **(D)** ascospores, **(E, F)** ascus tip showing apical chamber as seen by interference contrast **(E)** or phase contrast **(F)**, **(G)** ascospores. *Scale bars*: A, B = 0.5 mm, C = 100 µm, D–I = 10 µm.

5.1.2.2. Botryosphaeria rhodina

Botryosphaeria rhodina is characterized by abundant aerial and fast-growing mycelium, with conidia that are sub-ovoid, thick-walled, aseptate, and initially hyaline, which remains for a long time and finally becomes dark brown and one-septate.

5.1.2.3. Botryosphaeria obtusa

Botryosphaeria obtusa produces dark-green colonies but sparse mycelia. For this fungus, single, small, dark-brown, and black pycnidia develop in seven-days-old cultures, and conidia are initially hyaline and aseptate, becoming dark-brown with age.

5.1.2.4. Botryosphaeria ribis

Botryosphaeria ribis causes gummosis disease on plum. The conidia (14.2–26.8 × 4.3–7.2 µm) are single celled, hyaline, and spindle shaped. Asci (105–135 × 12.5–15.5 µm) are hyaline, clavate, and bitunicate. Ascospores (18–22 × 7.0–8.2 µm) are hyaline and spindle shaped or fusoid (Fig. 5.14).

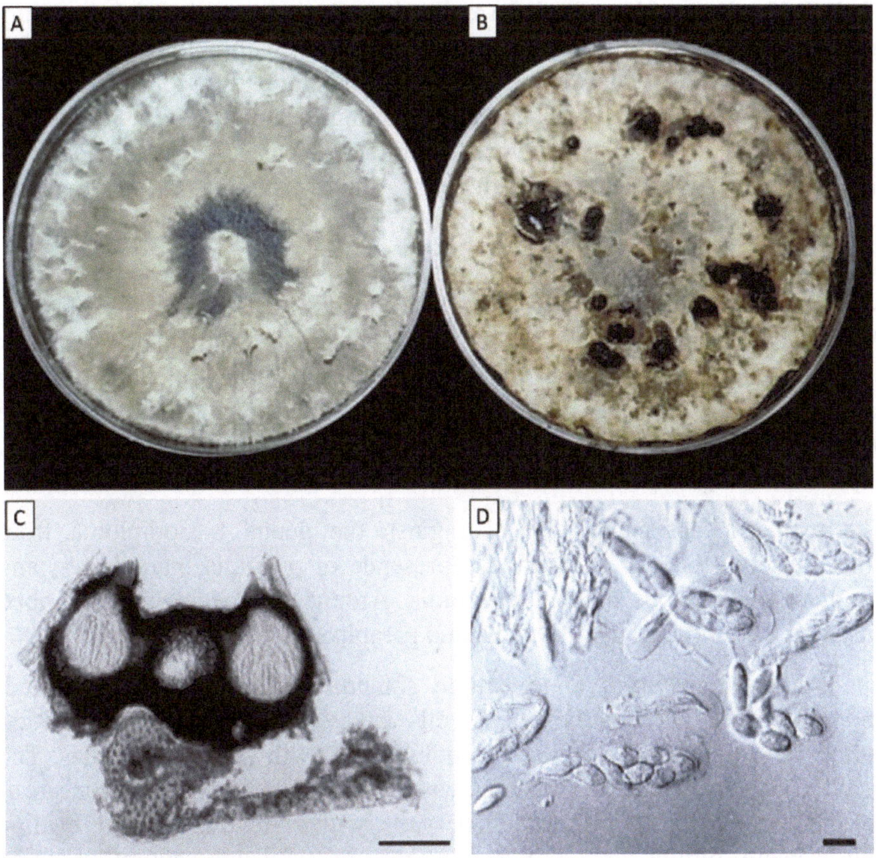

Fig. 5.14: Isolate of the fungus *Botryosphaeria ribis* on 3% malt agar after 10 days of fungal growth **(A)**, forms dark fruiting picnidial bodies with pycnidiospores **(B)**, Pseudothecia **(C)** (*Bar* = 140 μm), and asci and ascospores **(D)** (*Bar* = 20 μm).

5.1.3. Lasiodiplodia

Lasiodiplodia species (*L. theobromae, L. pseudotheobromae, and L. caatinguensis*) causes cashew gummosis.

5.1.3.1. Lasiodiplodia theobromae

It forms a gray and dense aerial mycelium on the culture medium. At first, hyaline immature conidia are developed, which later turn brown (25.1 × 12.0 μm on average), septate, and shows striation after 10 days. After exposure of the petri-plates to UV light (320 nm) for 14 days, solitary pycnidia are produced by the fungus.

The colonies in the culture medium are moderately dense, with aerial mycelium initially white, then turning to gray-olive color within seven days, and gradually acquiring a black color. The minimum temperature for the growth of *L. theobromae* is 15 °C, 28 °C optimum minimum, and 40 °C maximum. The sporulation of the fungus is favored by photoperiods of more than 16 h of light exposure, which allows for the formation of pycnidia. On the contrary, exposure to less than 4 h of daily light over a period of 23 days inhibits the sporulation of the fungus.

Mycelium is immersed or superficial, branched, septate, and dark brown. Ascomata are eustromatic, dark brown to black, uniloculate with a thick pseudoparenchymatic wall, ostiolate, embedded in the substrate, and partially erumpent at maturity. Pseudoparaphyses are hyaline and septate. Asci are bitunicate with thick endotunica and well-developed apical chamber, clavate, stipitate, and eight spores are present in it.

The main characteristics that distinguish the genus Lasiodiplodia from other closely related genera are the presence of pycnidia, paraphyses, and longitudinal striations in mature conidia. Around 20 species are described based on the morphology of conidia and paraphyses.

The conidiophores are hyaline, simple, sometimes septate, rarely branched, and cylindrical. The conidiogenous cells are hyaline, thick-walled, smooth, cylindrical to sub-obpyriforms, and holoblastic, with one or two rings. The conidia are sub-ovoids to ellipsoidal, with apex rounded, narrow to truncate at the base, wider mid-upper third, thick walls, with a granular content, hyaline and aseptate, turning dark brown once mature, with one septum, showing deposits of melanin on the interior surface of the wall longitudinally disposed, which gives a striated appearance with a size of 21.5–31.5 × 13–17 μm and one portion of 1.9 length and width.

Ascocarp is a dark brown to black aggregate, with a thick dark brown wall and hyaline internal layers, of 250–400 μm in diameter. The asci is bitunicate, stipitate, with eight spores, of 90–120 μm longitude. The ascospores are biseriated, hyaline, aseptate, with a size 30–35 × 11–14 μm. Ascospores are irregularly biseriate, initially hyaline, becomes dark brown, and aseptate. Conidiomata are stromatic, immersed or superficial, separate, or aggregated and confluent, globose, dark brown, uni- or multi-locular; walls of dark brown, thick-walled textura angularis, paler and thinner-walled towards the conidiogenous region, often with dark brown superficial hyphae over the surface. Ostioles are central, single, and papillate (Fig. 5.15).

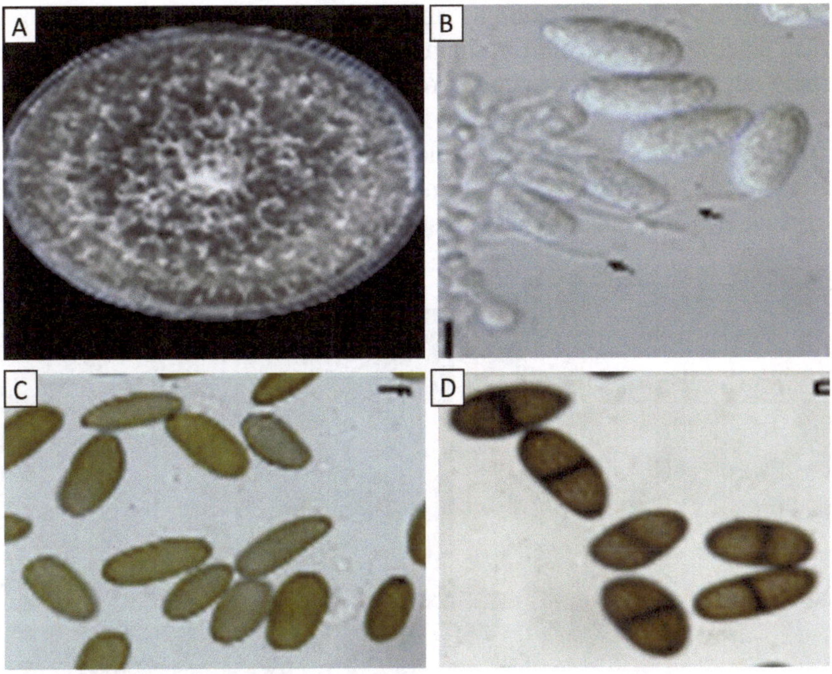

Fig. 5.15: *Lasiodiplodia theobromae.* **(A)** Colony morphology of 21-day-old *L. theobromae culture.* **(B)** conidiogenous cells and young hyaline and thick-walled *L. theobromae* conidia, non-septate paraphyses are indicated by black arrows, **(C, D).** Dark-brown mature *L. theobromae* conidia with longitudinal striations and central septum. *Scale bar* = 10 µm.

5.2. Collar Rot

5.2.1. Sclerotium

Sclerotium rolfsii produces abundant white mycelium on infected plants and in culture. Advancing mycelium and colonies often grow in a distinctive fan-shaped pattern, and the coarse hyphal strands may have a somewhat ropy appearance. In culture, mycelium appears smooth at first, but some cultures may develop aerial mycelia that cover all or part of the culture after a few days. The fungus produces at least two types of hyphae: large-diameter (5–9 µm) main branch hyphae and smaller-diameter (2–4 µm) sub-branch hyphae. Cells are hyaline, with thin cell walls and sparse cross walls. The main branch hyphae may have clamp connections on each side of the septum. In agar plate culture, sclerotia are not formed until the mycelium covers the plate.

In vitro or *in vivo*, sclerotia begin as small tufts of white mycelium that form spherical sclerotia 0.5–1.5 mm in diameter. Sclerotia darken as they mature, becoming tan to dark brown in color. Young sclerotia often exude droplets of clear to pale yellowish fluids. Mature sclerotia are hard, slightly pitted, and have a distinct rind. Although most sclerotia are spherical, some are slightly flattened or coalesce with others to form an irregular sclerotium. *S. rolfsii* does not form asexual fruiting structures or spores (Fig. 5.16).

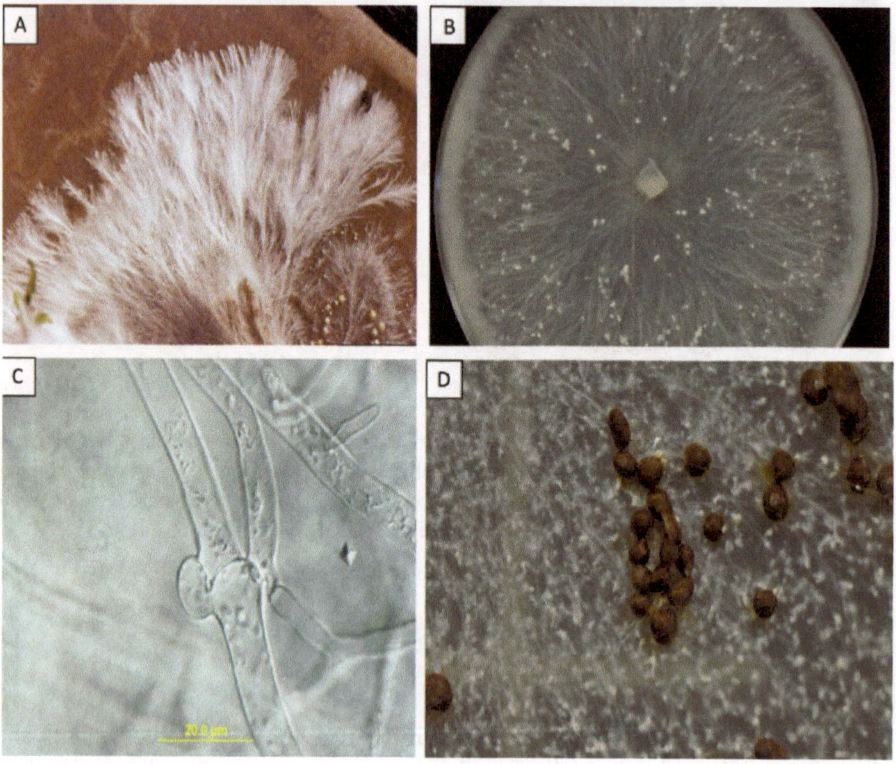

Fig. 5.16: *Sclerotium rolfsii.* **(A)** young ropy mycelial culture of *Sclerotium rolfsii* on quarter-strength potato dextrose agar (1/4 PDA), **(B)** Formation of young sclerotial bodies on the mycelial mat, **(C)** clamp connection in culture, **(D)** mature *Sclerotia* growing on potato dextrose agar (PDA).

5.2.2. *Athelia*

5.2.2.1. *Athelia rolfsii*

It is the perfect stage of *S. rolfsii.* Colonies develop within three days of incubation at 28° C. Hyphae on PDA are cottony white and spread to the periphery of the Petri dishes in a radial fashion. It has a well separated mycelium with branched hyphae. Clamp connections and septate hyphae are also present. *A. rolfsii* produces basidia on an exposed hymenium and basidia produces four haploid basidiospores. The appressed hymenium develops in small, thin, irregular patches. The clavate basidia are 4–6 μm × 7–14 μm and basidiospores are hyaline, 1.0–5 μm × 5–12 μm.

5.2.3. *Aspergillus*

5.2.3.1. Aspergillus niger

Colonies on potato dextrose agar at 25° C is initially white, which quickly becomes black with conidial production. The reverse is pale yellow and growth may produce radial fissures in the agar. In MEA (malt extract agar) media incubation for seven days is required at 25 and 37 °C producing slightly brown colonies and smooth-walled conidia. In Czapek Yeast Agar media after five days of incubation at 25 and 37 °C, it produces black colonies with wooly smooth-walled colonies of conidia.

Generally, they have a cottony appearance; initially white to yellow and then turns black. Made up of felt-like conidiophores. The reverse is white to yellow. In microscope, the conidial heads are radiate with conidiogenous cells biseriate. Conidia of *Aspergillus niger* are produced on conidiophores (Fig. 5.17).

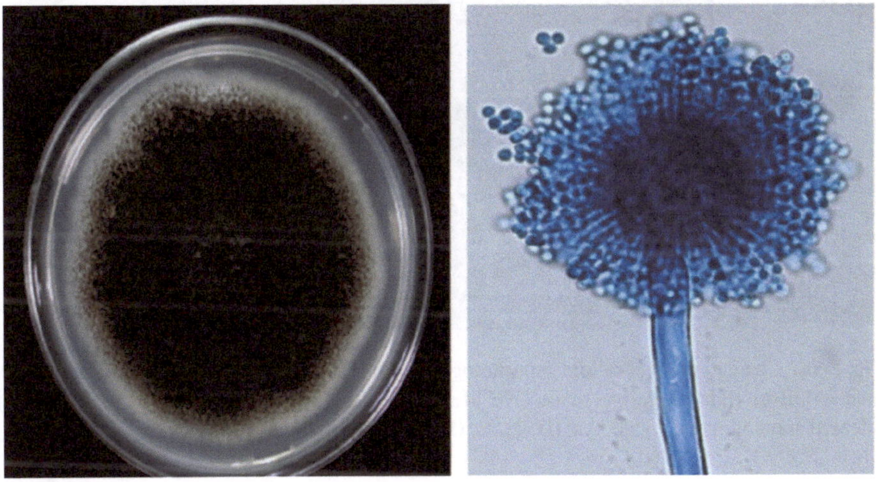

Fig. 5.17: Fungal colony characteristics of *Aspergillus niger* on PDA, and structure of *Aspergillus niger* under microscope.

5.2.4. Phytophthora

5.2.4.1. Phytophthra cactorum

Phytophthora cactorum causes collar rot of apple. It is homothallic and produces sexual oospores that can survive for many years in the soil.

Colonies of *P. cactorum* are whitish, fluffy, or smooth on PDA. Zoosporangia are oval to elongate (lemon shaped) with papillae. Terminal or intercalary chlamydospores are present. Antheridia and oogonia, as well oospores, are formed in large numbers (Fig. 5.18).

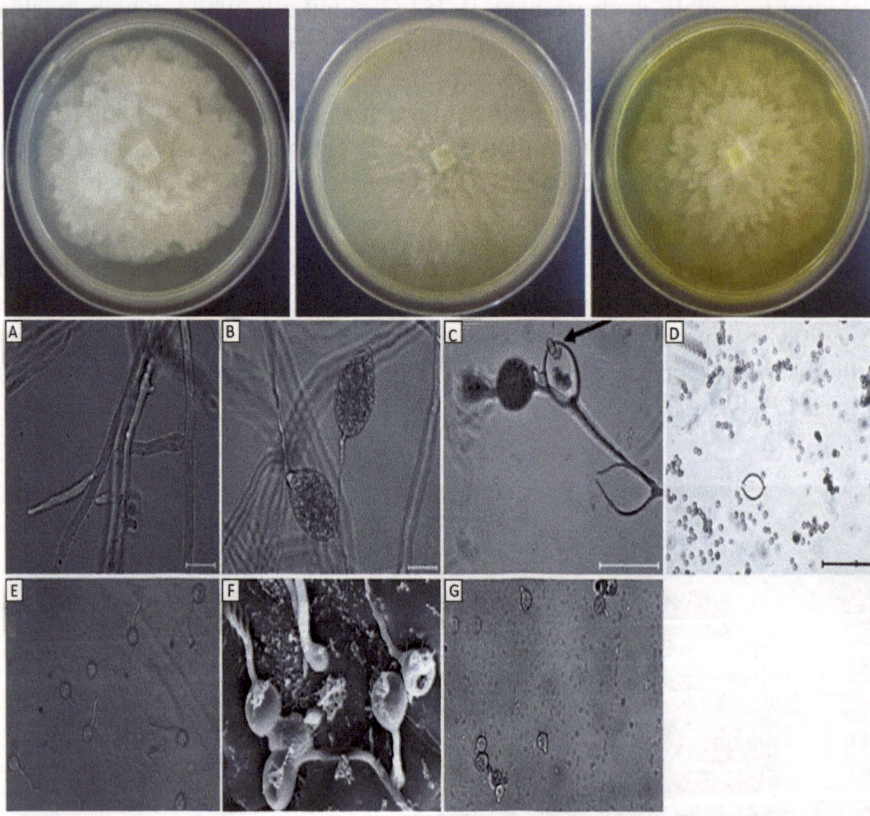

Fig. 5.18: Characteristic coralloid morphology of *P. cactorum* on PDA, V-8 Agar, and MEA media (left to right) and five successive life cycle stages. **(A)** mycelia, **(B)** sporangia, **(C)** released sporangia and zoospores, **(D)** cysts, **(E–G)** germinating cysts.

5.2.4.2. Phytophthora cambivora

It causes collar rot of apple. Colonies of *P. cambivora* are whitish, fluffy with dense aerial hypha. Colonies on V-8, PDA, and MEA media are with no distinctive pattern. Minimum temperature for fungal growth is 2° C, optimum temperature range is 22–24° C, and maximum temperature is 32° C.

Asexual phase

Zoosporangia are non-papillate, persistent, ovoid, obpyriform, or ellipsoid (55–65 µm × 40–45 µm), exhibit internal and often nested proliferation, and originate on simple, unbranched sporangiophores. Hyphal swellings are rarely produced and are coralloid with irregular shapes. Chlamydospores are absent.

Sexual phase

Heterothallic fungi. Oogonia has characteristic bullate wall projections. Antheridia are amphygynous, showing one or two cells when paired with the opposite mating type. Oospores are plerotic.

Most typical characters

Phytophthora cambivora is heterothallic, forming oogonia with characteristic warty, or bullate walls, and two-celled antheridia (Fig 5.19 – 5.20).

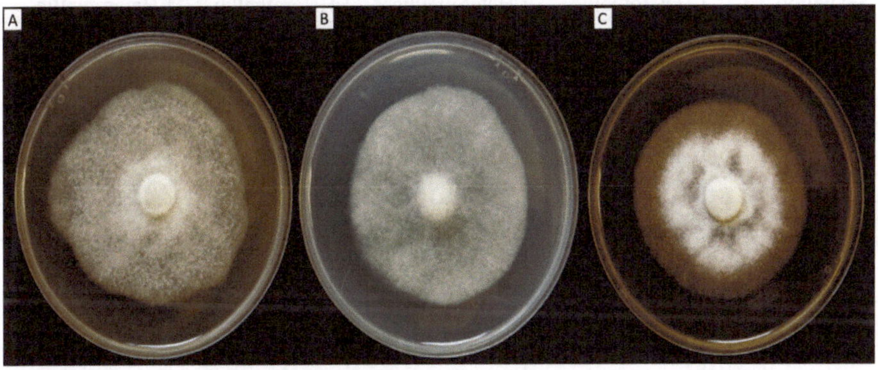

Fig. 5.19: *Phytophthora cambivora* colonies grown for seven days on **(A)** V-8 Agar, **(B)** potato dextrose agar, and **(C)** malt extract agar media.
Source: Photo courtesy by Clinton Greub, Krysta Jennings, and Leandra Knight, USDA-APHIS-PPQ.

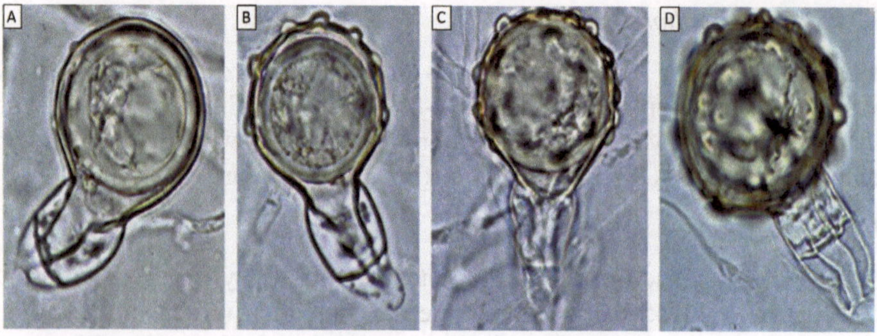

Fig. 5.20: *Phytophthora cambivora* sexual phase: **(A–D)** morphology of the oogonia, antheridia, and oospores after pairing with opposite type.
Source: Photo courtesy by G. Abad, USDA-APHIS-PPQ.

Other *Phytophthora* species causing collar rot in other crops including pomegranate (in Turkey) include *Phytophthora nicotianae, P. palmivora, P. parvispora* and *P. inundata*. These are differentiated on various morphological characteristics of sporangia, hyphal swelling, and others (Table 5.1, Fig. 5.21).

Table 5.1: Comparison of morphological characteristics of different *Phytophthora* spp.

Characteristics	***Phytophthora nicotianae***	***P. palmivora***	***P. parvispora***	***P. inundata***
Sexuality	Heterothallic	Heterothallic	Heterothallic	Heterothallic
Sporangia shape	Usually ovoid and spherical, sometimes obpyriform, rarely ellipsoid	Usually ovoid and obpyriform, sometimes spherical and ellipsoid, rarely limoniform	Usually ovoid, ellipsoid, and barreled, sometimes obpyriform, rarely pyriform	Ovoid and obpyriform
Papillae	Yes	Yes	No	No
Total range (μm)	30.2–69.3 × 24.2–58.8	32.6–66.6 × 21.7–44.2	34.9–82.6 × 19.5–47.4	41.4–79.8 ×31.0–61.1
Length and breadth mean (μm)	49.4 × 37.8	46.4 × 32.7	52.2 × 32.8	59.3 × 45.5
Length and breadth ratio	1.31	1.42	1.59	1.30
Internal proliferation and nesting	No	No	Yes	Yes
Caducity	Mostly non-caducous rarely caduceus with short pedicel	Caducous with short pedicel	Non-caducous	Non-caducous

Characteristics	*Phytophthora nicotianae*	*P. palmivora*	*P. parvispora*	*P. inundata*
Chlamydospore	Yes (36.5 µm diameter)	Yes (36.5 µm diameter)	Yes (rarely)	Yes
Hyphal swellings	No	No	Yes	Yes
Maximum temperature	>35° C	>35° C	>35° C	>35° C
Growth colony pattern	No	No	Chrysanthemum on MEA	Rosette on PDA and MEA

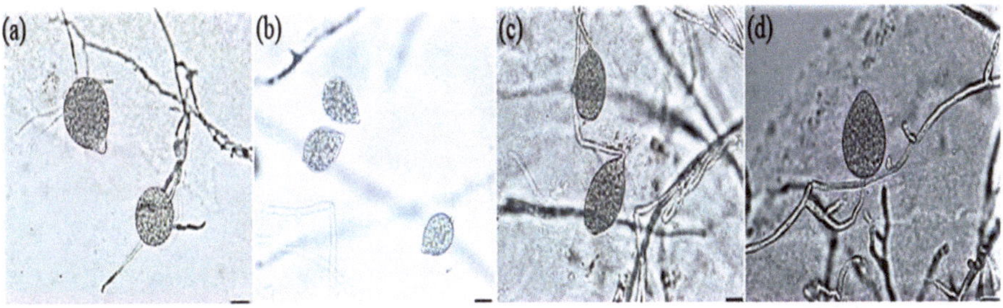

1. Sporangium in: (A) *P. nicotianae*, **(B)** *P. palmivora*, **(C)** *P. parvispora*, **(D)** *P. inundata*.

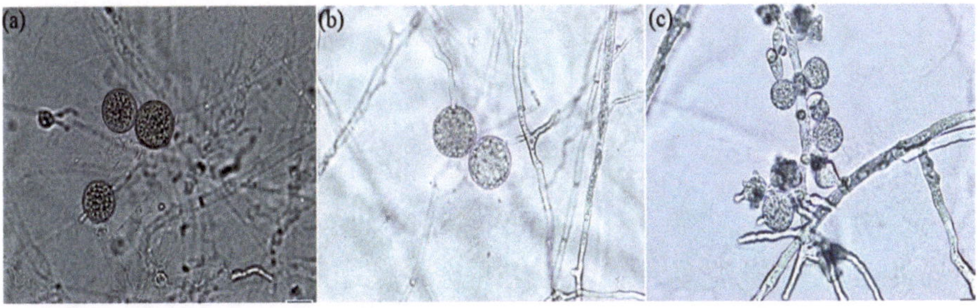

2. Chlamydospores in: (A). *P. nicotianae* **(B)**. *P. palmivora* **(C)**. *P. inundata*.

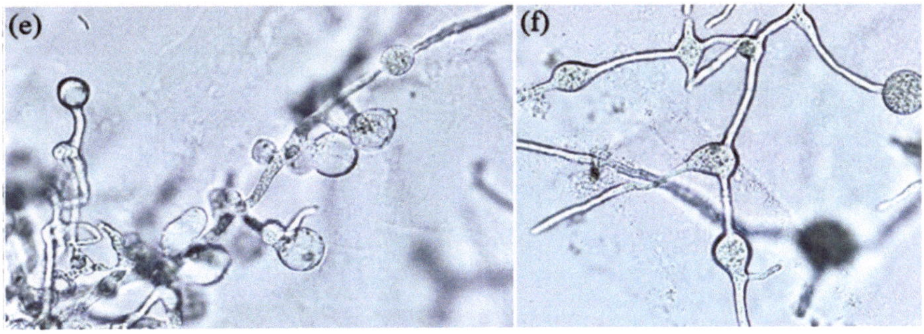

3. Hyphal swellings in: (E). *P. parvispora* and **(F)**. *P. inundata*. *Bar* = 20 µm.

Fig. 5.21: Sporangium, chlamydospore, and hyphal swelling in species of phytophthora.

6

Identification of Fungi Causing Foliar Diseases

6.1. Leaf Spot, Blight, and Blast

Leaf spots are irregular or round blemishes found on either side or both sides of the leaves. A typical spot is "zonal," meaning it has a definite edge and a darker border. When lots of spots are present, they can grow together to cover the larger areas of the leaves. Different fungi are responsible for the leaf spots and sometimes a particular fungus causes a peculiar type of leaf spot and includes the fungi.

6.1.1. Alternaria

Alternaria spp. have septate, brown hyphae. Conidiophores are also septate and brown. These bear simple or branched large conidia (7–10 × 23–34 μm) which have both transverse and longitudinal septations. Under the microscope, these spores will appear larger with a dark appearance.

The fungal colonies on the growth medium are fast-growing, black to olivaceous-black or grayish (Fig. 6.1).

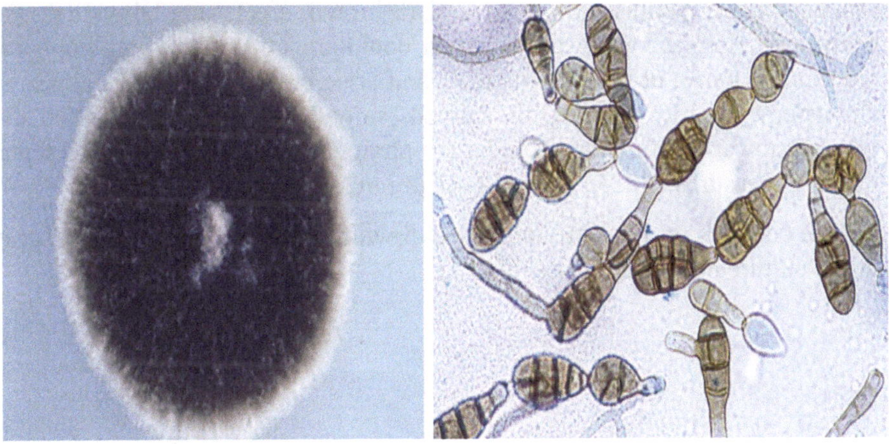

Fig. 6.1: Fungal growth of *Alternaria* on growth media and microscopic spores.

6.1.2. Phytophthora

Phytophthora particularly *phytophthora infestans* causing leaf spots produces microscopic, asexual spores called sporangia. These sporangia are hyaline (clear), lemon-shaped, and 20–40 µm long. Sporangia are formed on specialized branches called sporangiophores and the cytoplasm in the sporangia divides to develop many zoospores which emerge from each sporangium to develop further infection loci (Fig. 6.2).

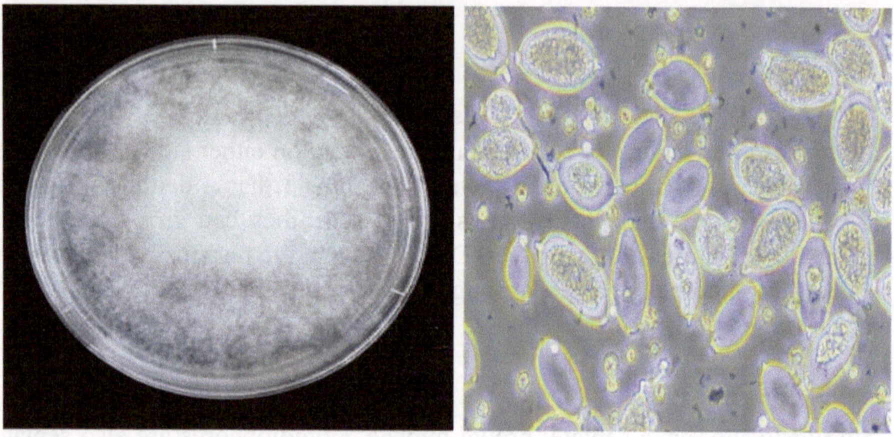

Fig. 6.2: Fungal growth of *Phytophthora* on growth medium and sporangia in microscope.

6.1.3. Helminthosporium

Helminthosporium produces hyphae, conidiophores, and conidia. Hyphae are septate. Conidiophores are brown to dark brown, erect, parallel-walled, and ceasing to elongate when the terminal conidium is formed. Conidiophores are septate, loose, or dense, brownish, and arise in groups. Conidia appeared elongate, cylindrical to slightly curved, smooth, and mostly rounded but tapering toward both ends. Conidia are phragmospores showing 6–10 septa. Conidia measures 67–155 µm × 14–21 µm in size.

Fungal colonies on PDA are initially dull whitish when young, and turn gray as the culture becomes older (Fig. 6.3).

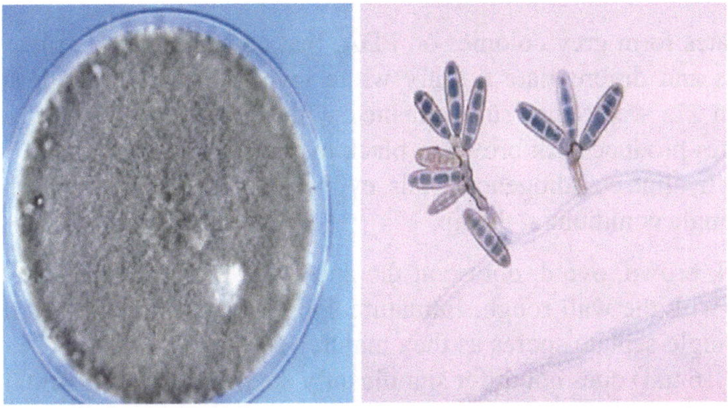

Fig. 6.3: Fungal growth of *Helminthosporium* on growth media and spores in microscope.

6.1.4. Curvularia

The fungal hyphae are septate, brown, with brown conidiophores, and conidia. Conidiophores are simple or branched and are bent at the points where the conidia originate in sympodial geniculate growth. The conidia (8–14 × 21–35 µm), which are also called poroconidia, are straight or pyriform, brown, multi-septate, and have dark basal protuberant hila. The conidia of *Curvularia lunata* have three septa and four cells, while those of *Curvularia geniculata* mostly have four septa and five cells.

Curvularia produces rapidly growing, wooly colonies on potato dextrose agar at 25° C. Initially, the color of the colony is white to pinkish gray and turns to olive brown or black as the colony matures (Fig. 6.4).

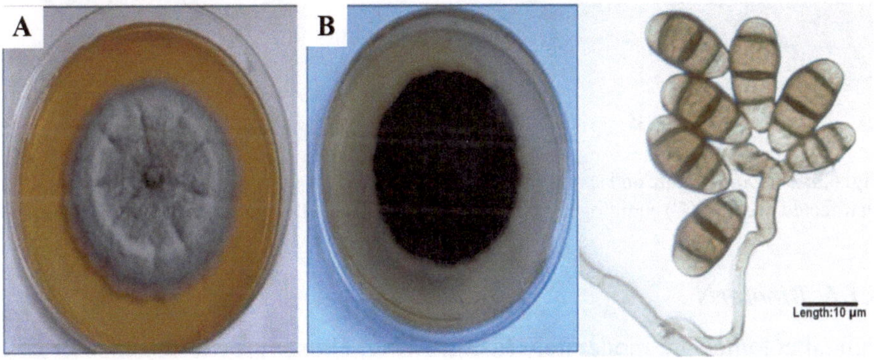

Fig. 6.4: Fungal growth of *Curvularia* on growth medium (A and B) and spores in microscope.

6.1.5. Diplodia

The fungal isolates form grey colonies on PDA, that covers the entire plates within five days and differentiate a fluffy white mycelium within 10 days. When grown on 2% water-agar supplemented with sterilized pine needles, the fungal isolates produce dark brown to black pycnidia after 14–20 days of incubation, with hyaline conidiogenous cells, cylindrical, thin-walled, smooth, and produce a single conidium at the tip.

Conidia are dark brown, ovoid, obtuse at the apex, and truncate or rounded at the base and with the wall rough. Immature spores are aseptate and form darker brown, single-septate spores as they mature. The size of conidia (24.4 ± 0.33 × 9.8 ± 0.16 μm) does not differ significantly among the isolates and fit with those reported for *Diplodia seriata.*

Diplodia seriata produce dark brown to black pycnidia after 14–20 days of incubation with hyaline conidiogenous cells, cylindrical, thin-walled, smooth, and produce a single conidium at the tip.

In *Diplodia maydis* mycelial surfaces are smooth. Slime-like material produced in liquid media appeared as a finely laced net, randomly appearing throughout the mycelia with bead-like structures present along the net. A fine netting is also observed interspersed among the spores inside the pycnidia obtained from oats (Fig. 6.5).

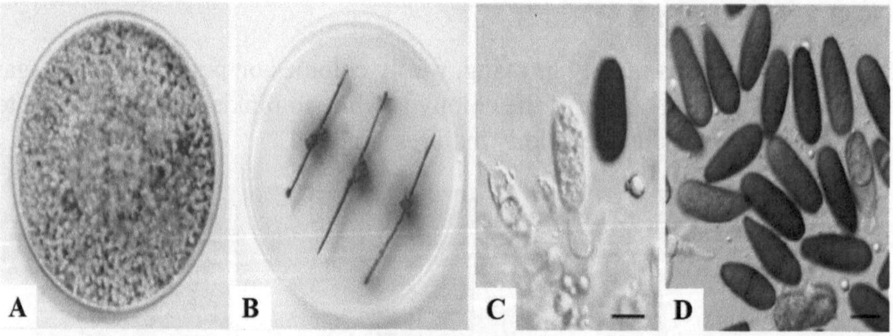

Fig. 6.5: *Diplodia* culture and spores: **(A)** fungal growth on growth medium, **(B)** pycnidia grown on infected needles, **(C)** conidiogenous cell with conidia, and **(D)** *Diplodia* spores in microscope.

6.1.6. Bipolaris

Fungal colonies are moderately fast-growing, effuse, grey to blackish brown, suede-like to floccose with a black reverse. Microscopic morphology shows sympodial development of hyaline to deep olivaceous, pigmented, pseudo-septate conidia on a geniculate or zig-zag rachis. Conidia are mostly curved, canoe-shaped, fusoid, or obclavate, rarely straight, 2–14 pseudo-septate (usually more than six), germinating only from the ends (bipolar).

Cochliobolus (1934) is the oldest name of *Bipolaris* (1959) and is more frequently used by plant pathologists (Fig. 6.6).

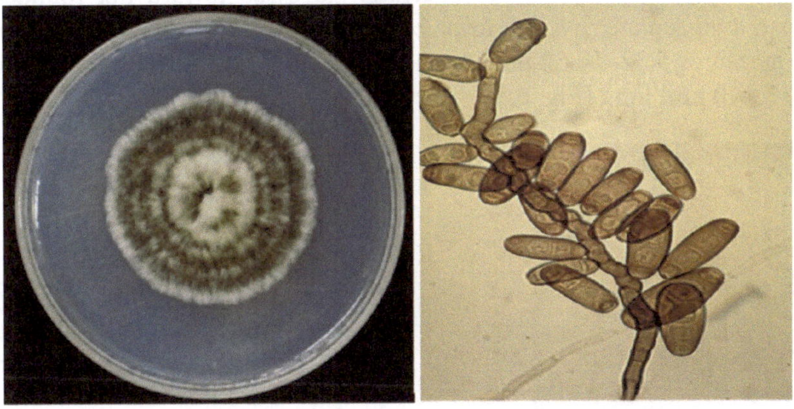

Fig. 6.6: Fungal growth of *Bipolaris* on growth medium and spores in microscope.

6.1.7. Stemphylium

Colonies of *Stemphylium* grow rapidly and velvety to cottony in texture, gray, brown, or brownish black in color. The septate hyphae are pale brown to brown in color. *Stemphylium* species have dark conidiophores that are swollen at their tips and give rise to single round or oval muriform conidia that are constricted at their central septum. These conidia have transverse and vertical septations (muriform conidia) and there is a typical constriction at the central septum (Fig. 6.7).

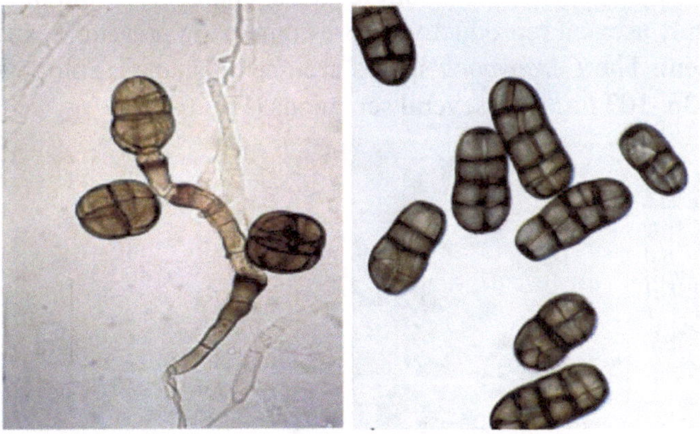

Fig. 6.7: Fungal spores of *Stemphylium*.

6.1.8. *Pyricularia*

The fungal culture color varies from grayish black to dark jet-black, with smooth to irregular margin.

Conidia are hyaline, pyriform to obclavate, narrowed toward tip, rounded at the base, two-septate, hyaline to pale brown, with a distinct basal hilum. Conidia measures 5.32–10.64 µm (average 8.24 µm) wide × 15.96–29.26 µm (average 25.40 µm) long (Fig. 6.8).

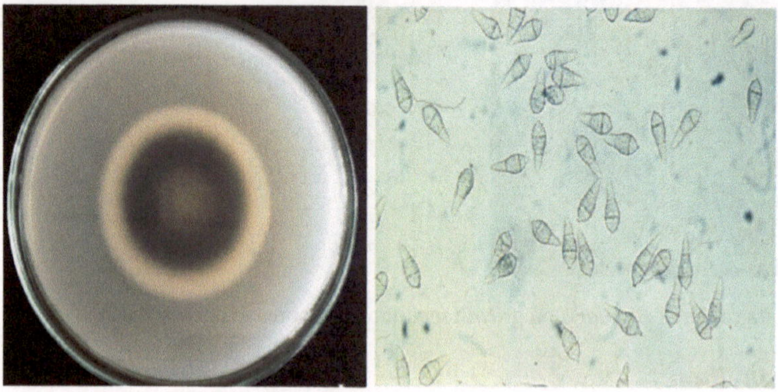

Fig. 6.8: Fungal growth of *Pyricularia* on growth media and spores in microscope.

6.1.9. *Cercospora*

Mycelium is well-developed, branched, septate, slender, intercellular, and brown colored. Conidiophores are septate, and appear as dark-colored structures coming out in tufts from stomata or ruptured epidermis. Conidia develop on distinct geniculate (knee like) bends structures of conidiophore. On liberation from conidiophore, each conidium leaves a small scar at the place of its attachment. Only asexual reproductive spore conidia are present. Sexual reproduction is absent. The *Cercospora* spores are needle-shaped, colorless, and measure 2–3 × 36–107 µm with several septations (Fig. 6.9).

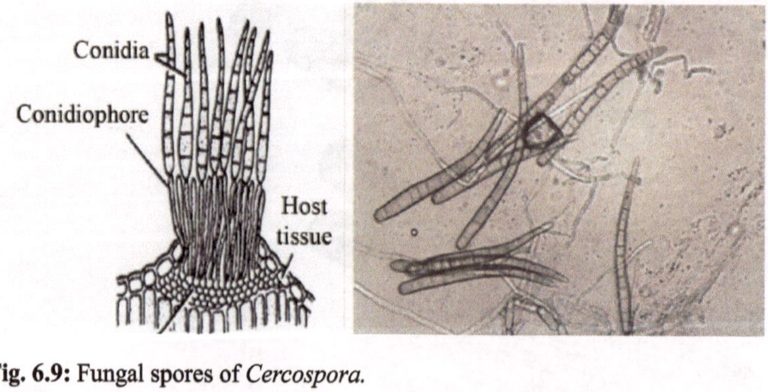

Fig. 6.9: Fungal spores of *Cercospora.*

6.1.10. *Colletotrichum*

The fruiting body of a fungus is a cushion shaped acervulus and seen below the epidermis or cuticle with dark setae. Setae is septate, stout at base and pointed at tip, dark brown, long, present in the periphery or in between the conidiophores. Conidiophores are simple, short, hyaline, non-septate, short cylindrical, falcate or crescent-shaped.

The conidia are mostly cylindrical or sickle shaped with rounded ends, hyaline, single-celled. They appear pinkish when in masses. The conidia measure 10–15.5 × 4–6.2 μm.

The fungal isolates in Petri plates are mainly whitish, grayish, or cream which appears as cottony and velvety mycelia on the top side of the culture and grayish creamy with concentric zonation on the reverse side (Fig. 6.10).

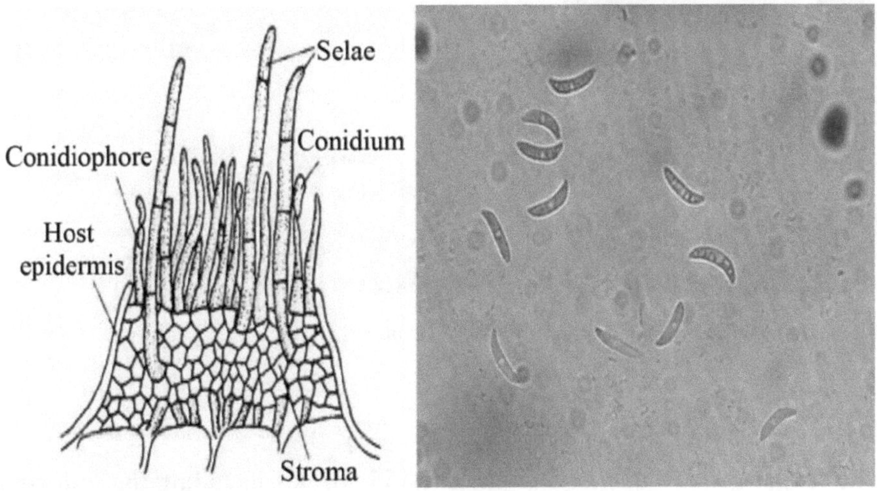

Fig. 6.10: Fungal fruiting body acervulus and spores of *Colletotrichum.*

6.1.11. *Myrothecium*

On culture plates, the fungal growth appears as white colonies with minute dark green droplets of sporodochia spread all over. Fruiting bodies are either synnema or sporodochia formed from closely compacted conidiophores, arising from a stroma bearing a mass of slimy green to black spores which become hard on drying. Hyaline or darkened setae are present or absent; the setae arise from the basal stroma.

Conidiophores are macronematous (morphologically different from vegetative hyphae), closely packed to form sporodochia, and branched repeatedly. The ultimate branches bear conidiogenous cells (phialides) in whorls; they are hyaline, olivaceous, or slightly darkened.

Conidiogenous cells are monophialidic and discrete, cylindrical, elevated or subulate, hyaline, or darkened at the apex.

Conidia are aggregated in dark green or black slimy masses; they are simple, cylindrical, navicular, limoniform, or broadly ellipsoidal. They often have a projecting truncate base; they are hyaline to pale olive, smooth or straightly marked, and non-septate. The conidia measure 5–8 × 1–2.5 µm (Fig. 6.11).

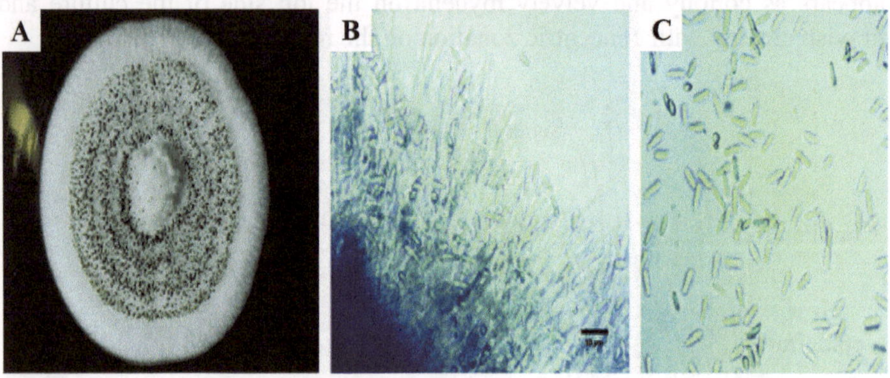

Fig. 6.11: Fungal growth and spores of *Myrothecium.*

6.1.12. Cochliobolus

The fungal genus *Cochliobolus* includes 55 species including the following plant pathogenic species: *C. carbonum*, *C. heterostrophus*, *C. miyabeanus*, *C. sativus*, and *C. lunatus*.

The genus *Cochliobolus* is distinguished by the presence of dark to black ascomata with a unilocular, globose pseudothecium and a short, cylindrical neck. Ascomata also bear hyphae and conidiophores, and asci are bitunicate, 2–8 spored, and cylindrical to obclavate. Ascomata are black, globose to ellipsoidal, and setae are present over the upper half of the wall mixed with conidiophores. Pseudoparaphyses are filiform, hyaline, septate, and branched. The shapes of asci are cylindrical to clavate, short-stalked, straight to slightly curved, having 1–8 spores, and vestigial bitunicate.

In the asexual fungal stage, the conidia are straight to moderately curved, occasionally cylindrical but usually broad in the middle and tapering toward the rounded ends, distoseptate, and 36–100 × 12–18 μm. The surface of the conidia is often granulose and the hilum is inconspicuous. Conidia are produced from the apex of an unbranched conidiophore. Generally, the conidiophore arises singly or in small groups that are straight or flexuous, mid to dark brown, smooth, septate, cylindrical, and up to 250 μm long, 5–8 μm thick. Variation in mycotoxin production by *Cochliobolus* species is used to distinguish some taxa and these mycotoxins are host-specific and non-host-specific. HC toxin produced by *C. carbonum* race 1 and T toxin produced by *C. heterostrophus* are host-specific toxins while ophiobolins produced by *C. miyabeanus* and carbotoxin produced by *C. carbonum* are non-host-specific toxins.

Previously, many species of *Cochliobolus* were placed in *Ophiobolus*. Drechsler found that graminicolous forms of *Helminthosporium* produced teleomorphs with characteristic features of greater ascus and ascospore width as well as helicoid ascospores with bipolar germination, which deviates from previously described characters of *Ophiobolus*. To accommodate these species, he introduced the new genus *Cochliobolus* and transferred several species to *Cochliobolus*. Another genus *Pseudocochliobolus* was separated from *Cochliobolus* based on the presence of stromatic tissue below the ascomata and the degree of ascospore coiling.

The sexual states of *Cochliobolus* are only found associated with *Bipolaris* and *Curvularia*. *Bipolaris* and *Curvularia* share some morphological similarities and cannot be easily distinguished by any distinct taxonomic criteria. Although few morphological differences were reported, the conidia of *Bipolaris* are distoseptate while the conidia of *Curvularia* are bigger and euseptate. Some scientists believe that *Bipolaris* and *Curvularia* are synonymous, however, there is an existing debate regarding this issue. The different species of fungi belonging to *Cochliobolus* with asexual states in *Bipolaris* and *Curvularia* have undergone frequent name changes as a result of refinement to the taxonomy that has resulted in some confusion. The perfect stage of this fungus was first named *C. carbonum* by Nelson in 1959. The imperfect stage was named as *Bipolaris zeicola* in 1959. The commonly used synonyms of *Cochliobolus carbonum* were *Helminthosporium zeicola* (1930), *Helminthosporium carbonum* (1944), *Drechslera zeiocola* (1966), and *Drechslera carbonum* (1984). Several phylogenetic studies of *Bipolaris* and *Curvularia* showed that *Bipolaris* is not monophyletic and some *Bipolaris* species are nested within *Curvularia* (Fig. 6.12).

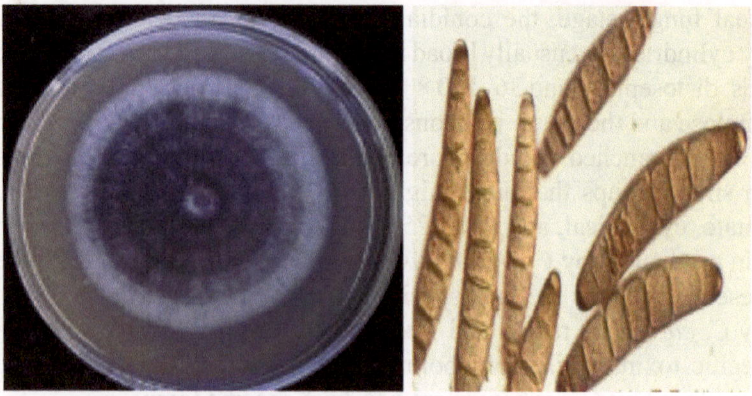

Fig. 6.12: Culture of *Cochliobolus* with asexual spores.

6.1.13. Cladosporium

Cladosporium or *cladosporioides* has sparse, unbranched or rarely branched, darkly pigmented hyphae that are typically not constricted at the septa. Mature conidiophores are treelike and comprise many long, branched chains of conidia. The presence of shield-shaped conidia, a distinct hilum, and chains of conidia that readily dis-articulate are characteristic of the genus *Cladosporium*. *Cladosporium or cladosporioides* reproduces asexually and because no teleomorph has been identified, it is considered an exclusively anamorphic species. The shape of the spores varies from spherical, to barrel-like and to lemon-shaped.

Colonies are olive green to olive brown and appear velvety or powdery. On a potato dextrose agar (PDA) medium, colonies are olive-grey to dull green, velvety and tufted. The fungal spore measures 2.7–13.3 × 2.1–5.9 µm (Fig. 6.13).

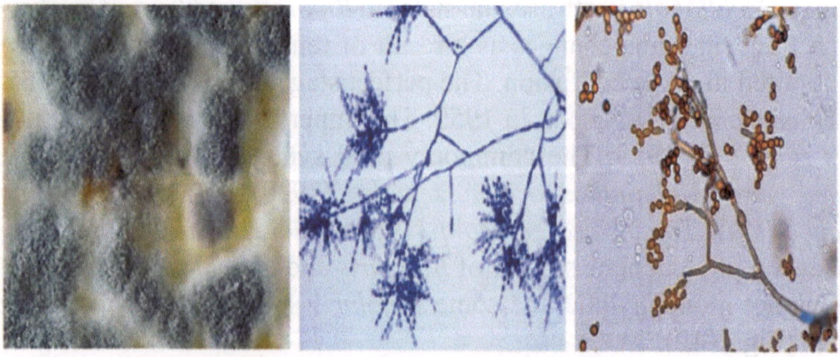

Fig. 6.13: Fungal colony of *Cladosporium* and fungal spore structures.

6.1.14. Ascochyta

The fungus produces hyaline to brown and septate mycelium. The asexual fruiting bodies pycnidia are spherical to sub globose with a prominent ostiole. Different species of *Ascochyta* vary in colony color from light brown to black, grey or dark brown on growth media. The size of pycnidia and pycnidiospores also varies significantly. Maximum size of pycnidia is reported from isolates AR 4 is 225 × 224 μm. The pycnidiospore or conidia are hyaline, oblong to oval-in-shape, straight or slightly bent at both ends, single celled or occasionally bi-celled and measures 6–12 × 4–6 μm (Fig. 6.14).

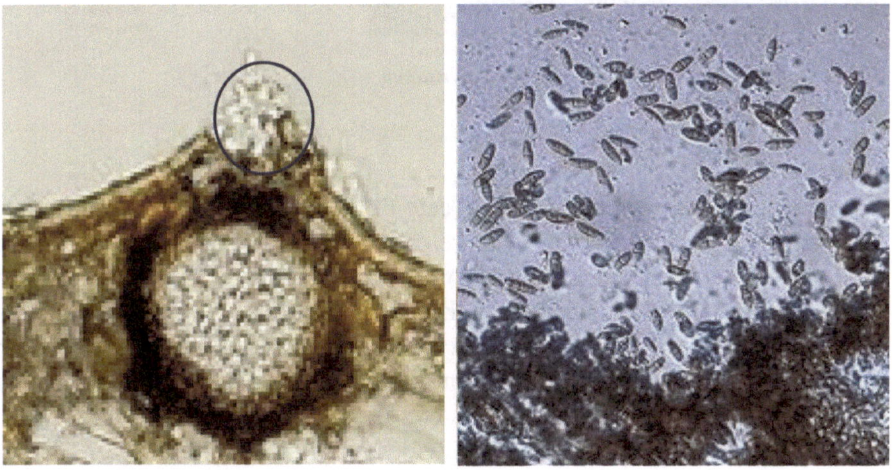

Fig. 6.14: Pycnidia of *Ascochyta* with pycnidiospores.

6.1.15. Exobasidium

Exobasidium genus of fungi in the family *Exobasidiaceae* has a widespread distribution, especially in northern temperate regions, and contains about 50 species. Host-specific fungal pathogens in the genus *Exobasidium* can cause reddening, distortion, blisters, and gall formation in leaf, flower, and stem tissue on many different plant species worldwide.

Exobasidium produce mature two celled basidiospores which easily dislodged from the sterigmata and are found on the surface of the blister. The basidiospore is one septate, as many as three septa have been seen in germinated spores. The spore size varies from 3–5 × 10–15 μm (Fig. 6.15).

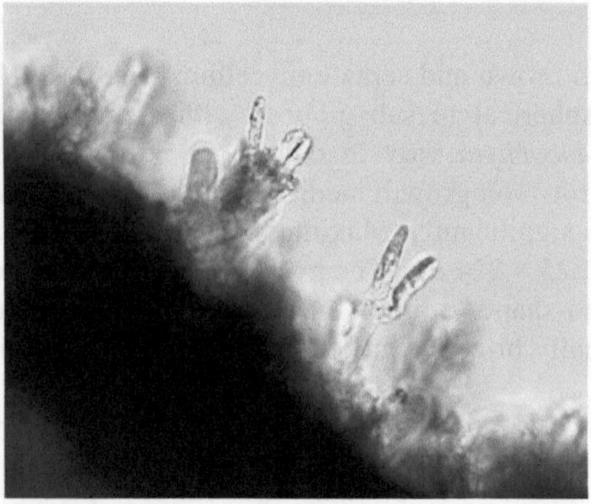

Fig. 6.15: Basidia with basidiospores of *Exobasidium.*

6.1.16. Venturia

Venturia genus belongs to an ascomycete fungus which produces ascospores as sexual spores in a sac like structure called ascus. The ascocarps appear in the form of pseudothecia. They are solitary and embedded into the host plant tissue. A pseudothecium has small dark hairs around its opening and contains pseudoparaphyses along with asci. Asci are elongated, sac-like structures, each of which contains eight ascospores in a linear arrangement. The ascospores are brown, two-celled spores, and have a characteristic "footprint" shape. Ascospores measure between 5–7 µm wide and 11–15 µm long.

The mycelium of *Venturia* is septate, and the nuclei are haploid in condition. *Venturia also* reproduces asexually by spores called conidia. Conidia are single-celled, uninucleate, and narrower at one end than the other. In mass, conidia appear brown or olive, but they are lighter when viewed individually under a microscope. Conidia measure between 6–12 µm wide and 12–22 µm long and are produced by specialized short hyphae called conidiophores. Conidiophores are formed on a dense mat of mycelia that pushes up through and ruptures the leaf cuticle. It is this mass of conidia and conidiophores that causes the velvety appearance of young scab lesions.

Conidia are produced nine to thirty days after initial leaf infection, depending upon temperature. They are disseminated by wind and by wind-driven rain. Both ascospores and conidia require a period of wetness to germinate. The germination hypha penetrates the cuticle and establishes a new infection. There can be many cycles of conidial production and infection within a single growing season (Fig. 6.16).

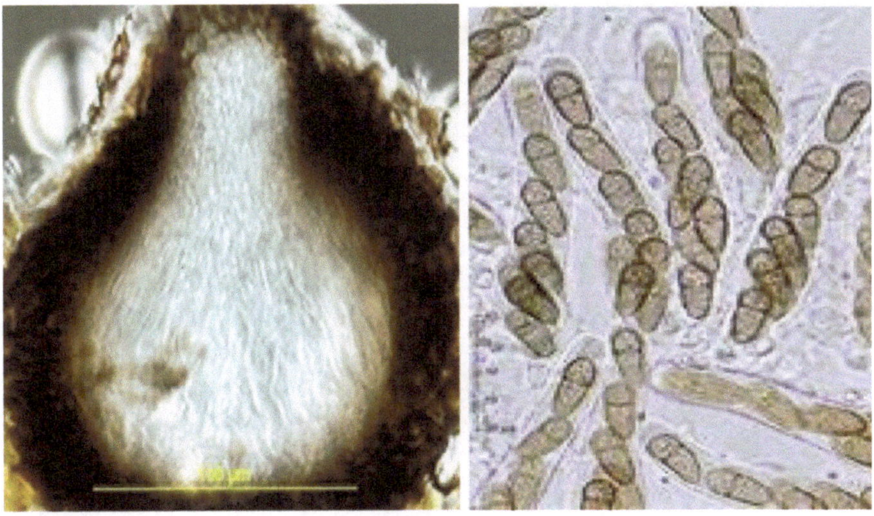

Fig. 6.16: Ascocarp and ascospore of *Venturia* sp.

6.1.17. Physoderma

Physoderma is a genus of chytrid fungi. Members of *Physoderma* are obligate parasites of pteridophytes and angiosperms. There are approximately 80 species within this genus. *Physoderma* species are characterized as having both a monocentric thallus and an endobiotic polycentric thallus. Resting spores germinate in the spring to produce zoospores that will infect the host. The initial infection gives rise to monocentric, epibiotic zoosporangium anchored with endobiotic rhizoids confined to a single host cell. The zoosporangium has been characterized as *Rhizidium* or *Phlyctochytrium*-like; it usually has discharge papilla through which the zoospores are released. The liberated zoospores infect new host cells, and in this fashion, an infection can go through several generations. As well, the sporangia are internally proliferous; that is, they can produce a second round of zoospores after releasing the first one. In late spring and summer, the zoospores will begin to develop into an endobiotic polycentric thallus. This thallus is often extensive, infecting many host cells, with highly branched, fine rhizoids. These rhizoids can bear intercalary cells, which many be 1–2 septa. The endobiotic thallus gives rise to large, thick-walled, dark-colored resting spores that take the shape of the host cell. It appears the resting spores are formed from the intercalary cells. These resting spores will over winter and germinate in the spring. There is a marked variability in resting spore size in the range of 14–39 × 25–59.7 μm. The pycnidia of the fungus are scattered, single, spherical, and dark brown. The mature spore wall consists of a single layer which is approximately 1–3 μm in thickness and light brown (Fig. 6.17).

Fig. 6.17: Resting spores of *Physoderma.*

6.1.18. Rhizoctonia

Rhizoctonia is a genus of fungi in the order *Cantharellales*. Species form thin, effused, corticioid basidiocarps (fruiting bodies), but are most frequently found in their sterile, anamorphic state where it is composed of hyphae and sclerotia (hyphal propagules). The hyphae are 4–15 µm wide and tend to branch at right angles. A septum near each hyphal branch and a slight constriction at the branch are diagnostic feature of the genus. It produces white to deep brown mycelium on an artificial medium (Fig. 6.18).

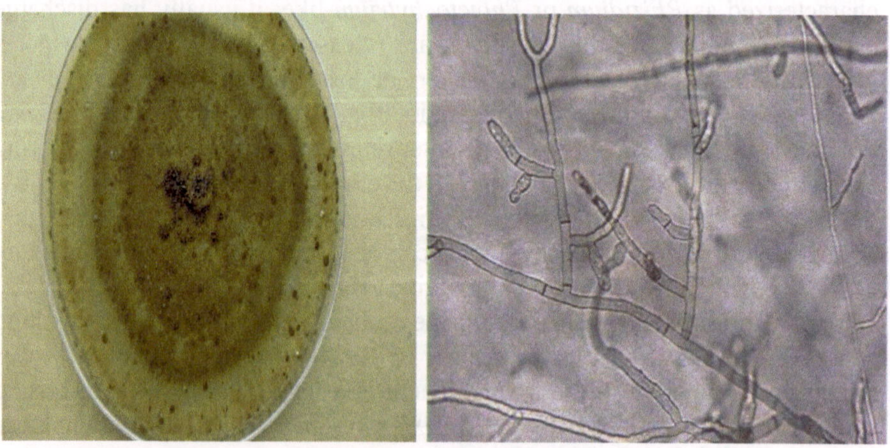

Fig. 6.18: Fungal colony on growth media and fungal structure.

6.1.19. Cephalosporium

Cephalosporium hyphae are fine, hyaline and produce mostly simple long slender phialides on which cylindrical or ellipsoidal conidia are formed in slimy bundles at the tips of the phialides. The conidia are usually one celled (i.e. ameroconidia), hyaline or pigmented, globose to cylindrical and mostly aggregated in slimy heads at the apex of each phialide (Fig. 6.19).

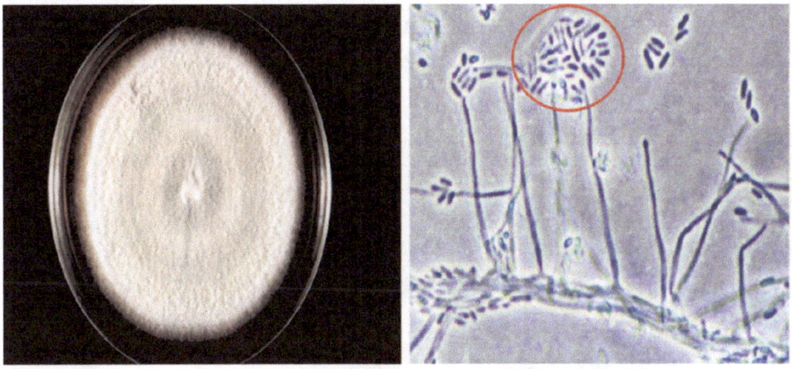

Fig. 6.19: Fungal colony of *Cephalosporium* and microscopic structure of fungi.

6.1.20. Didymella

In vitro, the fungal growth on an agar plate looks rough and undulated. When grown *in vitro* on agar, the fungus produces a white to olive-colored mycelium. In latter periods of growth, the mycelium is an olive to dark green or black color. Conidia are rounded, cylindrical, mono-septate, and hyaline. Conidia measure 6.4–13.6 µm in length and 3.69–4.68 µm in diameter. Black specks (perithecia and pycnidia) are formed on the infected parts or cankers (Fig. 6.20).

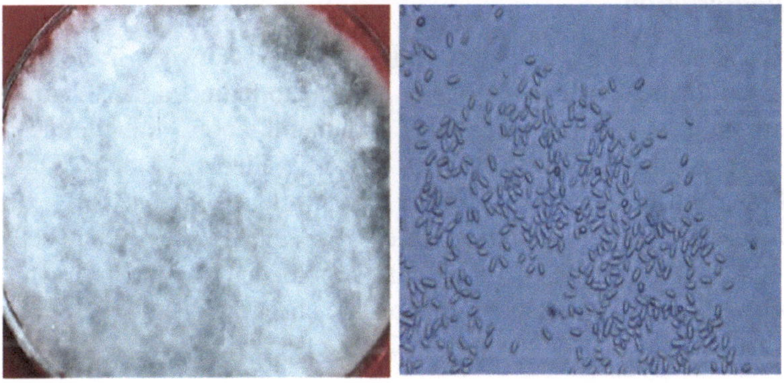

Fig. 6.20: Fungal culture and microscopic spores of *Didymella*.

6.1.21. Aureobasidium

Aureobasidium pullulans is a ubiquitous black, yeast-like fungus. It produces smooth, faint pink, yeast-like colonies covered with a slimy mass of spores. Older colonies change to black due to chlamydospore production. Hyphae are hyaline, smooth, and thin walled, with transverse septa. Primary conidia are hyaline, smooth, ellipsoidal, one-celled, and variable in shape and size, secondary conidia are smaller. Conidiophores are undifferentiated, intercalary, or terminal, or arising as short lateral branches. Endoconidia are produced in an intercalary cell and released into a neighboring empty cell (Fig. 6.21).

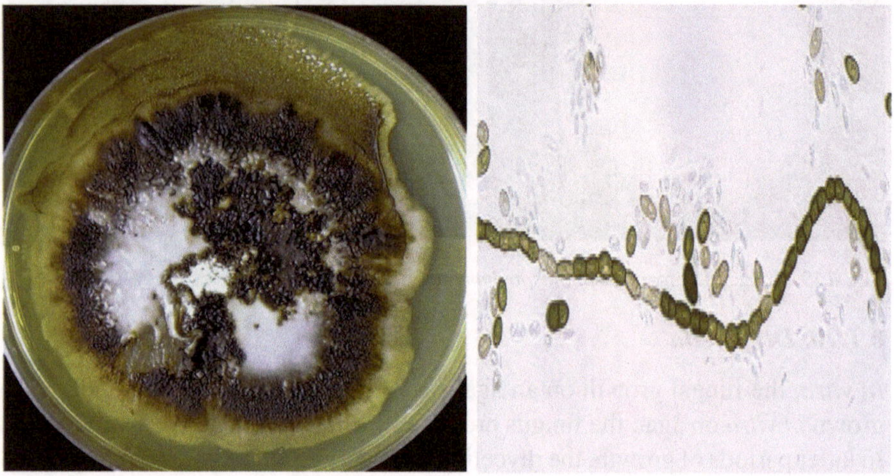

Fig. 6.21: Fungal culture and microscopic spores of *Aureobasidium.*

6.1.22. Hyalothyridium

Pycnidia epiphyllous, separate, sparse, serially aligned between host veins, at first immersed, then erumpent. These are globose to flattened flask-shaped, ostiolate, with thin walls (one cell thick) and membranous, subhyaline to pale brown then turning dark brown with thicker-walled (2–3 cells thick), a round ostiole, and a superficial portion. The wall is composed of extra angularis. Ostiole is well-defined and bordered by specialized centripetal cells. Conidiogenous cell holoblastic, hyaline, short, simple, determinate, intimately associated with pycnidial wall. Conidiophores are absent.

Conidia are large, solitary, ovate to oblong, 15–20 septate, muriform, becoming constricted at the septa at maturity under humid conditions, finally with cells often splitting apart, each capable of germinating (Fig. 6.22).

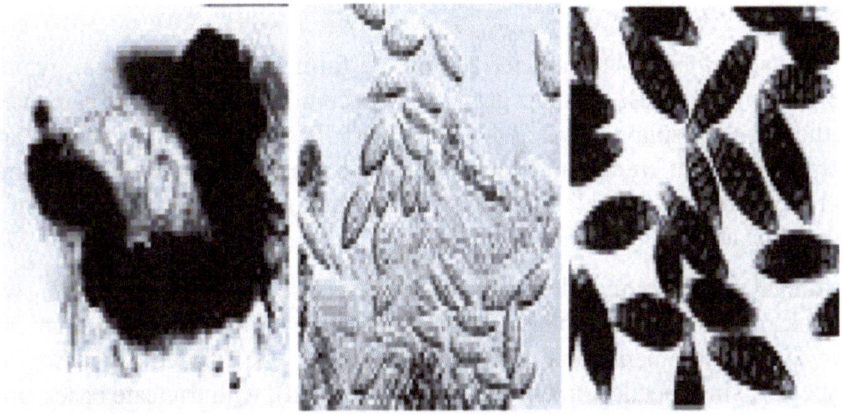

Fig. 6.22: Pycnidia and conidia of *Hyalothyridium* fungus.

6.1.23. Septoria

Conidia (spores) are produced within the pycnidia (fruiting structures). Pycnidia are dark, separate, globe-shaped with an ostiole (opening at the apex) from which conidia are extruded and erupt through the surface of the infected plant tissue. Conidia are produced on short conidiophores. They are clear, narrowly elongated to filiform, and several-celled. *Septoria* conidia may be confused with *Cercospora* conidia when seen alone. However, *Cercospora* conidia are produced on open condiophores, whereas *Septoria* conidia are produced within pycnidia.

Septoria produces globular, black pycnidia filled with very narrow and long conidia. The conidia are hyaline, straight, or slightly curved and one to five septate, they vary in length from 36 to 70 μm or longer and are 1.5–3 μm in diameter (Fig. 6.23).

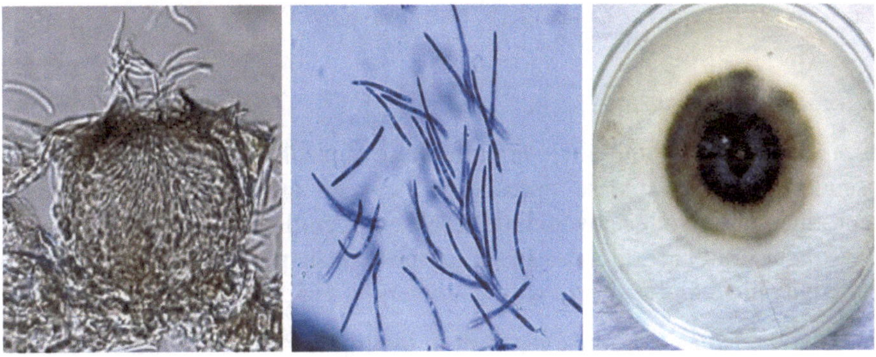

Fig. 6.23: Pycnidia, conidia and fungal culture growth of *Septoria*.

6.1.24. Graphium

The genus *Graphium* is characterized by the formation of synnemata, which consist of a more or less compact group of erect conidiophores that are cemented together, usually splaying out and bearing conidia at the apex. Synnemata are darkly pigmented, erect, and occur alone or in clusters. Conidia are hyaline, one-celled, smooth, sub-globose to ovoid, and are usually aggregated in slimy heads at the apex of the synnemata.

The fungal colonies on cultural medium are effuse, grey, olivaceous brown or black. It has septate hyphae, conidiophores are simple, long, and dark in color. The predominant conidia are hyaline but may be colored depending on the species, single-celled, allantoid (sausage-shaped), with truncate bases, and measured 4–7.5 µm × 1.3–2.2 µm (average 5.5 × 1.9 µm) (Fig. 6.24).

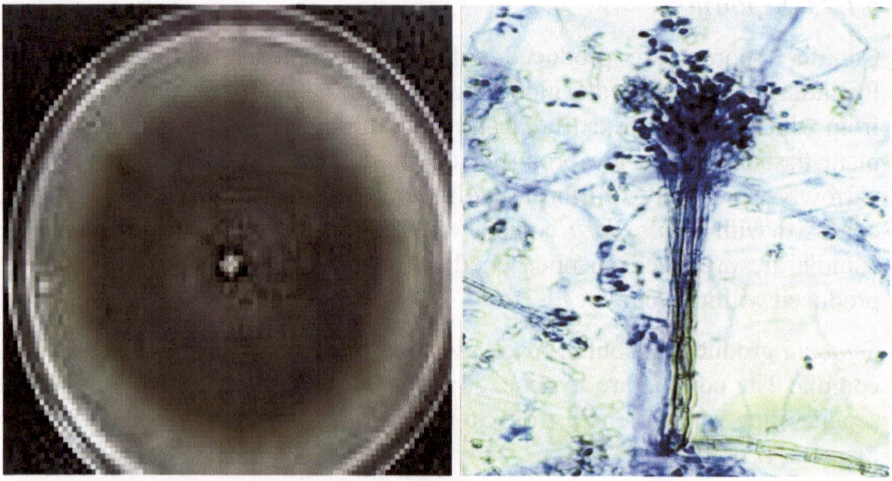

Fig. 6.24: Fungal culture and synnemata of *Graphium* sp.

6.1.25. Leptosphaeria

Leptosphaeria has both a teleomorph phase (sexual reproduction to generate pseudothecia that release ascospores) and an anamorph phase (asexual reproduction to produce pycnidia that release pycnidiospores). The pycnidia of the fungus are black, globose to sub-globose in shape. The conidia are single celled, hyaline, and fusiform with diameters of 4–5 × 1.5–2 µm.

Most of the isolates produce pigment in the liquid culture in variable color of brown to black (Fig. 6.25).

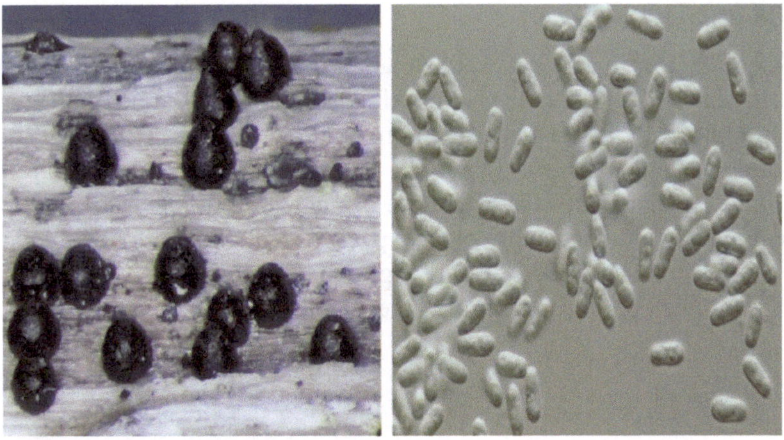

Fig. 6.25: Pycnidia and conidia of *Leptosphaeria.*

6.1.26. Leptothyrium

The pycnidia are depressed pulvinate, single or in groups, intra-or sub-epidermal (occasionally in the spongy parenchyma), of irregular shape, and 35–95 µm in diameter. The conidiophores are simple, septate, straight, or slightly curved, and 14–18 × 2–4 µm. The conidia are irregular, oblong, or ovate, sometimes curved, hyaline, continuous (but developing a septum at germination), and measure 8–13 × 2–3 µm. They germinate by germ-tubes from either or both ends. The typical plant pathogenic species is *Leptothyrium nervisedum* (Fig. 6.26).

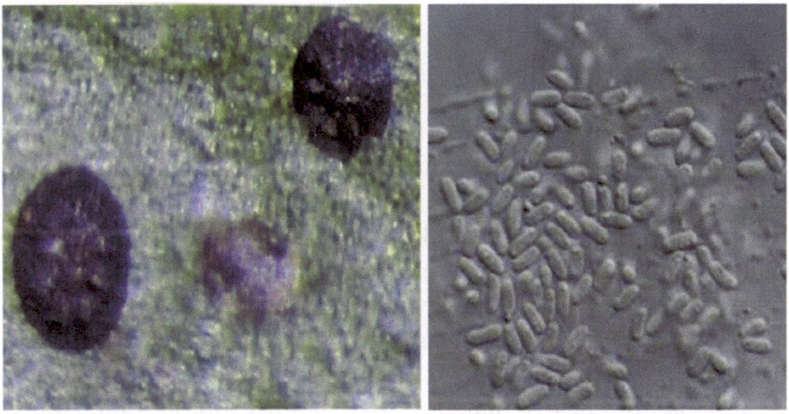

Fig. 6.26: Pycnidia of *Leptothyrium* and their conidia.

6.1.27. Ophiosphaerella

Ophiosphaerella is primarily a soil inhabitant with dark, flat brown infection mats of aggregated dark hyphae, approximately 60–400 μm in diameter, surrounding infested plant tissue are more often observed in nature. The fungi produce asci and ascospores in erumpent pseudothecia, which are rarely observed in nature. Pseudothecia are black, produced on stolons and roots of the infected plants, and are usually closely packed with a globose body and a thick neck. Ascospores are 35–180 μm long (depending on species), filiform, slightly twisted in a bundle, and parallel to each other (Fig. 6.27).

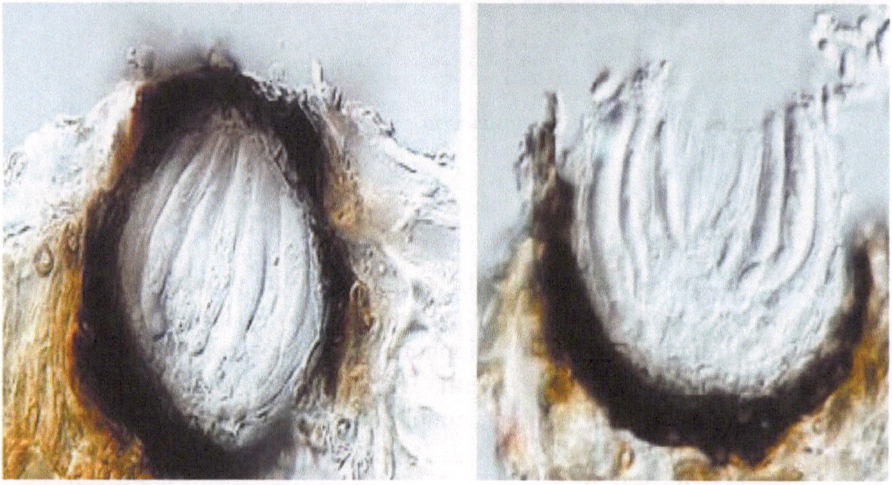

Fig. 6.27: Pseudothecia of *Ophiosphaerella*.

6.1.28. Paraphaeosphaeria

Stromata initially subepidermal, becoming erumpent at maturity, black, 270–390 (av. 336 μm) × 325–430 μm (av. 404 μm) of thick-walled pseudoparenchyma, ascomata often stromatic, solitary or several to many embedded in each stroma, walls of solitary ascoma of 9–10 layers of thick-walled, brown to dark brown scleroplectenchymatous cells, cells at apex of ascomata of smaller diameter, lacrimoid to narrowly pyriform, ostiolate, no periphyses observed. Ascospores cylindrical, broadly rounded at apex, tapering to a more narrowly rounded base, mostly 3 to 5 septate, constricted at penultimate septum, cell above this septum often slightly wider, uni- or bi-seriate within ascus, densely punctate, dark yellowish brown and conidia hyaline, bacilliform, ellipsoid (Fig. 6.28).

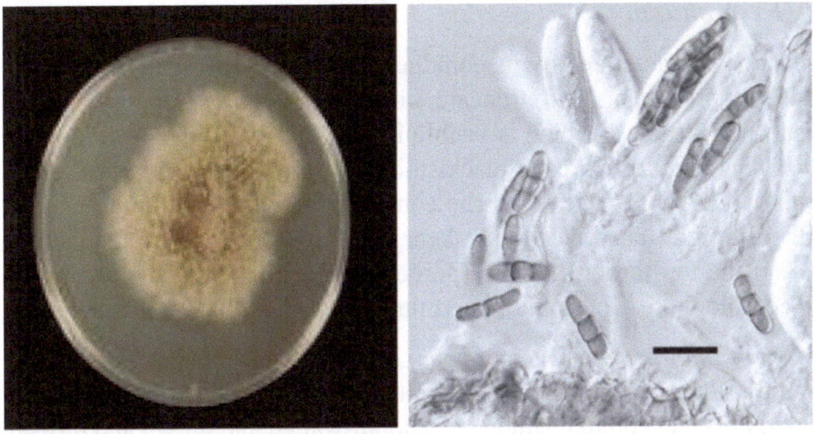

Fig. 6.28: Fungal culture of *Paraphaeosphaeria* and ascus with ascospores.

6.1.29. Setosphaeria

Setosphaeria (anamorph *Exserohilum*, formerly known as *Helminthosporium*) is the causal agent of northern corn leaf blight in maize. *Setosphaeria* produce mature conidia with a distinct protruding hilum. The conidia are either straight, curved or bent and the septum above the hilum is thickened and dark. The walls are typically roughened and brown to olive in color and with typically 7–9 septa (Fig. 6.29).

Fig. 6.29: Fungal culture and conidia of *Setosphaeria*.

6.1.30. Exserohilum

The genus is characterized by its conidia, which are ellipsoidal, distoseptate, and have a protruding and truncate hilum. *Exserohilum* spores have an inner cup-like structure which is visible in the basal cell. Fungal colonies of *Exserohilum* range from grey to blackish-brown color. The texture varies from suede-like to floccose in texture. The conidia from which the hilum extends, are either straight, curved, slightly bent or ellipsoidal to fusiform, and are formed on the top through a pore (poroconidia) on an elongated sympodial angled conidiophore. The conidial germination of *Exserohilum* is bipolar.

Exserohilum ongirostratum is characterized by larger conidia (up to 228 × 12–19 µm), with 6–16 distosepta centrally curved. In contrast, *E. mcginnisii* has slightly clavate conidia, which are smooth-walled and brown and measure 44–76 × 11–18 µm with 4–8 distosepta lacking darkened bands (Fig. 6.30).

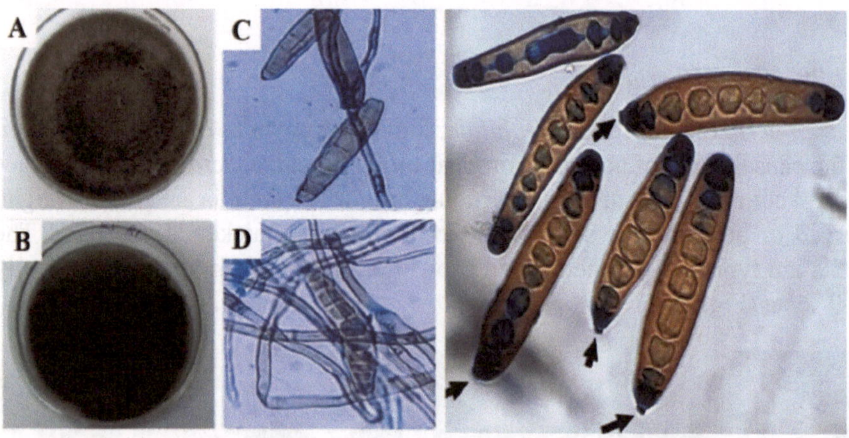

Fig. 6.30: Fungal culture **(A, B)**, and conidia **(C, D)** of *Exserohilum* with arrows showing the hilum.

6.1.31. Phaeosphaeria or *Parastagonospora*

The fungal colonies are round in shape. It has a white periphery with a grey central part. The reverse side of the colonies have a reddish-brown color. With age, the colonies darkened. On the above side, their color changes to grey-greenish with violet hue, while on the reverse side it is dark gray-black. The pigment released in the culture medium is yellowish-brown in the beginning, which becomes light brown.

The pycnidia of the fungus are scattered, single, spherical, and dark brown. *Phaeosphaeria* produce long, filiform, colorless, or colored, one to several celled conidia depending on the species (Fig. 6.31).

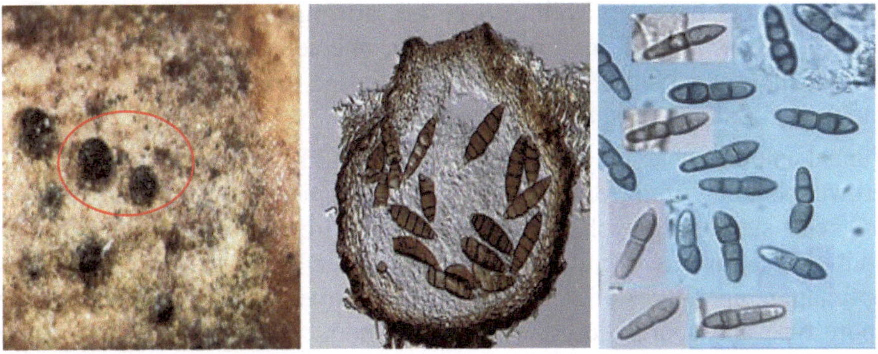

Fig. 6.31: Pycnidia and conidia of *Phaeosphaeria.*

6.1.32. Selenophoma

Maire (1906) erected the genus *Selenophoma* to cover pycnidial species with hyaline, falcate spores. On the basis of Maire's description, Sprague and Johnson (1940) concluded that certain fungi with non-septate, falcate spores borne in small globose pycnidia with coarse globose peridial cells were more logically assigned to *Selenophoma* than to *Septoria.*

The fungus grows readily in culture through dilution technique and make good and rapid growth on PDA producing a colony of 1.5–2 mm. in 2–3 days. Mycelium is submerged to sub-aerial, scanty, usually radial ashy-white.

The pycnidial formation in fungal culture is rapid and good at the end of 48 h and profuse at the end of seven days. They are in characteristic concentric rings, typically flask-shaped, morphologically similar to those from the host, but almost twice as big as on host (190–357 μm). Fresh conidia germinate very capriciously without developing any septation, as was observed in old spores.

Selenophoma eugeniae (obtained from *Eugenia jambolana*) is having globose to pitcher shape pycnidia with a protruding ostiole. The size of pycnidia is 98–182 × 98–210 μm, while the picnidiospores are 15–19 × 3 μm in size. The pycnidia do not have the neck.

Selenophoma kamatii sp. Nov. Kalani infection spots are indefinite, amphigenous, brick-red, Pycnidia dark brown to black, chiefly hypophyllous, scattered, profuse, flask-shaped, provided with a prominent neck opening through host tissue by a broad ostiole, subepidermal, carbonaceous, leathery, thick-walled, 92–148 × 182–260 μm. Neck dark-colored, 42–71 × 38–63 μm, ostiole broad, protruding out, 16.2 × 21 μm.

Conidiophores bulbous, hyaline, 1-celled, arising in tufts from a cushion-shaped, basal "mound" of tissue. Conidia one-celled, hyaline, allantoid to fusoid, slightly curved, 21–30 × 3.5 μm, released in a cirrus through ostiole. Incites defoliation in *Eugenia jambolana* Lam.

Fungal colonies obtained from conidia are yeast like in contrast to the hyphal growth obtained from pycnidial inoculum. Pycnidia and sclerotia are induced on media supporting hyphal growth and absent from conidial media. Growth and sporulation are significantly greater under near UV irradiation with optimum temperatures of 20 and 15° C, respectively. The medium determine spore type and sporulation intensity. Maximum sclerotial production occur under dark incubation at 15° C. *Selenophoma* produces both pycnidiospore and conidial inoculum respectively, on plates of Czapek Dox V-8 (CDV-8) agar and a modified Czapek Dox agar (MCDA) at 12° C under the cultural conditions (Fig. 6.32).

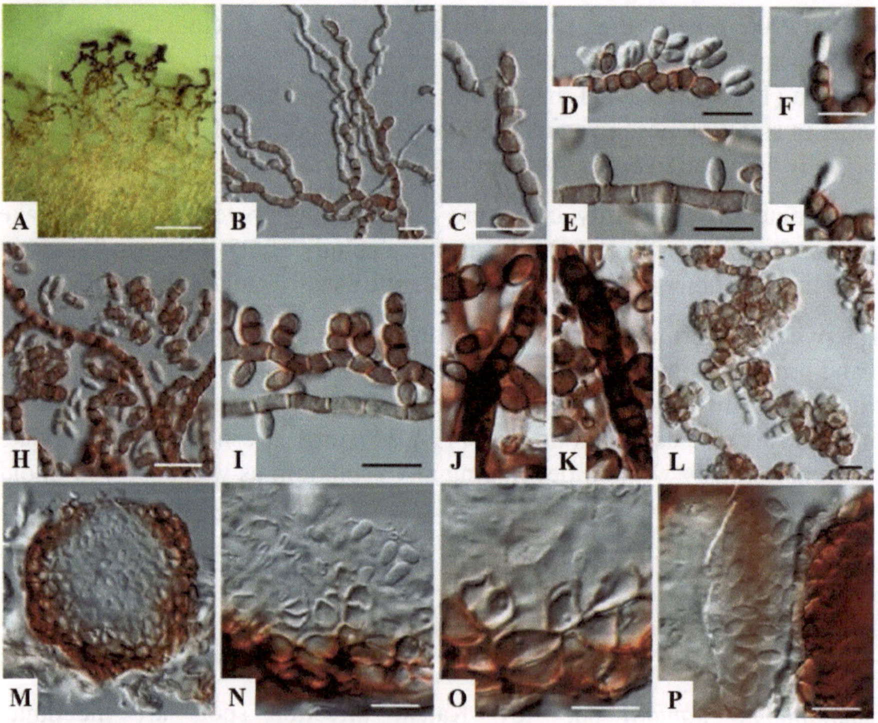

Fig. 6.32: Pycnidia and conidia of *Selenophoma.*

6.1.33. Gaeumannomyces

Gaeumannomyces is pathogenic on many members of the Poaceae family, but is most commonly a problem on wheat, barley, rye, oats and turfgrass.

This fungus produces a perithecial ascocarp within the host tissue composed of stromatal and host tissues. The ascospores are multiseptate, filliform and conidia may act as spermatia. The size of ascospore varies with f. sp. of the pathogen e.g. the size of ascospores in *G. graminis* var. *avenae* is 65–176 μm whereas in *G. gramins* var. *tritici (Ggt)* it ranges from 27 to 124 μm in length and in *Gaeumannomyces graminis* var. *graminis (Ggg)* it is 75–105 μm. The hyphae of Ggg is brown in color, septate, whereas in Ggt it produces chlamydospores under water stress conditions, which can induce formation of microsclerotia and chlamydospores in culture medium (Fig. 6.33).

Fig. 6.33: Perithecia and ascospores of *Gaeumannomyces*.

6.1.34. Phyllachora

Phyllachora is a perithecial ascomycete which causes a tar spot disease of maize. The fungus produces perithecia which are sub-globose, ostiolate, aggregated, or scattered, subepidermal. Paraphyses numerous, filiform, longer than asci. It produces pycnidia as asexual fruiting body and conidiophores branched at two or three levels, branches tapering, conidia filiform, hyaline, 10–15 μm, gradually tapering to apex (Fig. 6.34).

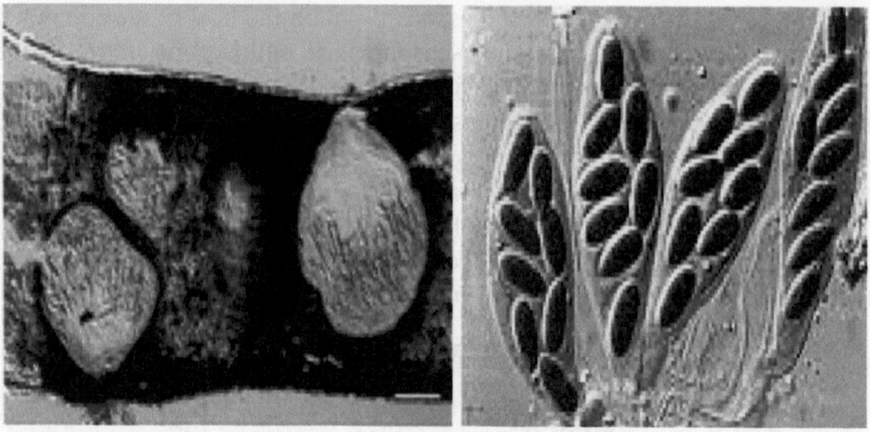

Fig. 6.34: Perithecia, asci and ascospores of *Phyllachora.*

6.1.35. *Monographella* or *Microdochium*

Mycelial colonies are flat, margin entire, slightly raised to umbonate center, pinkish white with reverse grayish orange on PDA media.

Monographella species are characterized by the production of perithecia immersed in leaf sheaths in natural substrates. In culture perithecia are superficial, globose, with clavate periphyses, show a peridium composed by isodiametric to subglobose cells of *textura angularis-epidermoidea*, and apically free paraphyses. Asci are oblong to clavate, with eight biseriate ascospores, and with a refractive, amyloid, flat, funnel-shaped apical ring. Ascospores are fusiform or oblong, hyaline, straight or slightly curved, and smooth. *Monographella* presently includes 11 species.

Microdochium and *Monographella* are important plant pathogens, particularly on grasses and cereals. In cold to temperate regions, "*Microdochium* patch," also known as "pink snow mold" is an economically important disease of wheat and barley, caused by *Microdochium nivale* and *M. majus*, and leaf-scald disease of rice by *Monographella albescens*. In Mexico, *Monographella maydis* on *Zea mays* produces a tar-spot disease complex of maize together with *Phyllachora maydis*. *Microdochium bolleyi* is known to produce root necrosis and the decay of grasses. *Monographella opuntiae* causes the brown spotting on *Opuntia*. *Microdochium tripsaci* is responsible for a systemic infection on *Tripsacum laxum*, while *M. sorghi* causes zonate leaf spots and decay on *Sorghum* species and other *Poaceae*, and *M. paspali* is known to produce the seashore paspalum disease of *Paspalum vaginatum* (Fig. 6.35).

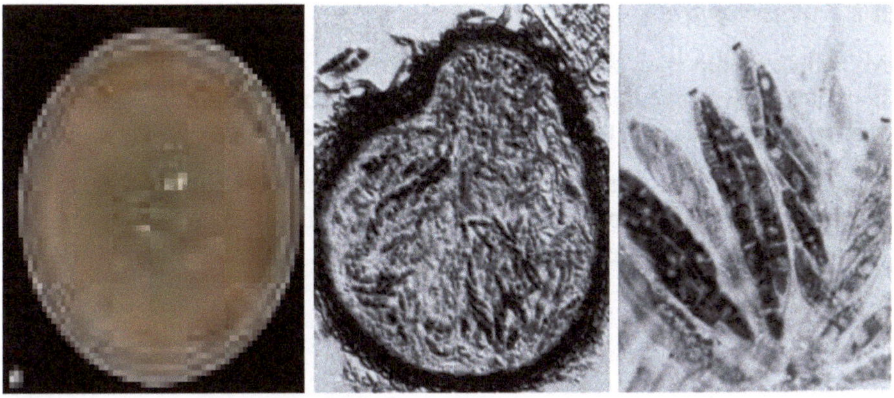

Fig. 6.35: Fungal culture, perithecia and ascospores of *Monographella.*

6.2. Downy Mildew

Downy mildew-causing fungal genera are differentiated on the basis of sporangiophore branching structure, sporangial shapes and their germination, and oospore structures. Different genera causing downy mildew are:

6.2.1. Peronospora

Sporangiophores are dichotomously branched 2–7 times at acute angles and tips of branches are curved and pointed bearing sporangia on sterigmata. Sporangium are hyaline, ovoid, non-papillate and always germinate by germ tube and sporangia behave like conidia (Fig. 6.36).

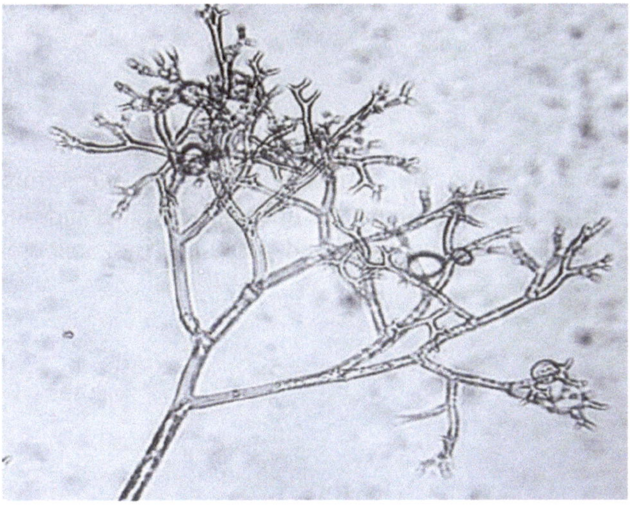

Fig. 6.36: Sporangiophore and sporangia of *Peronospora.*

6.2.2. *Plasmopara*

Mycelium composed of intercellular, colorless, aseptate hyphae 7–12 μm diameter, often irregularly shaped and swollen, bearing small, rounded vesicular haustoria, 4–10 μm diameter, formed predominantly in leaf tissues.

Sporangiophores are branched at right angles to the main axis at regular intervals. Monopodial branching is observed. Subsequent branches are 3–6, which end in blunt sterigmata of 3 in number. Sporangia are ovoid and germinate by zoospores (Fig. 6.37).

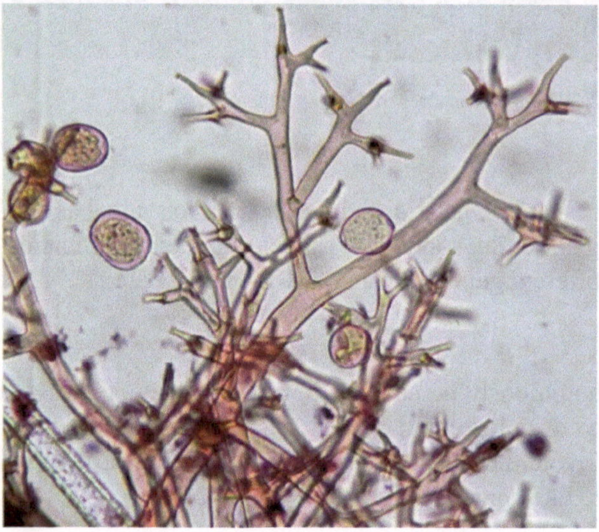

Fig. 6.37: Sporangiophore and sporangia of *Plasmopara.*

6.2.3. *Pseudoperonospora*

Sporangiophores are branched at acute angles with curved, blunt tips, bearing sporangia on sterigmata. Sterigmata are unequal (one big and one small). Sporangia are grayish, ovoid, papillate and germinate by zoospores (Fig. 6.38).

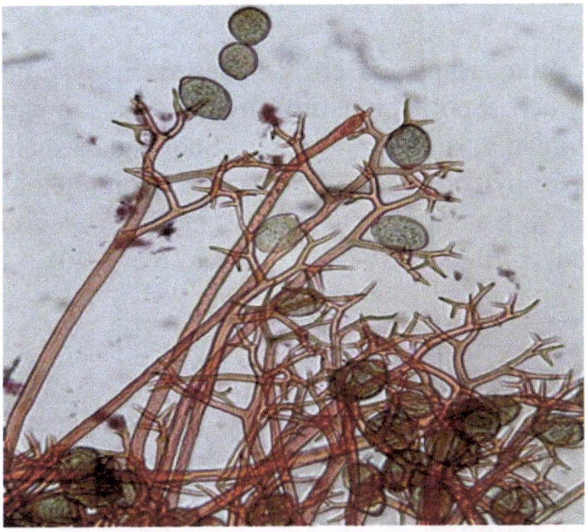

Fig. 6.38: Sporangiophore and sporangia of *Pseudoperonospora.*

6.2.4. Paraperonospora

Paraperonospora differ from *Peronospora* in having an uncolored dissemination unit and globose to pyriform haustoria and from *plasmopara* by the production of sporangia that germinate with a germ tube.

Paraperonospora differs from both of these genera in having sporangiophores that are widening toward the ramification. Sporangiophore are hyaline, straight, tree-like and mostly dichotomously to monopodially branched in 4 to 5 branchlets. Ultimate branchlets are straight to slightly curved and have truncate or obtuse tips. Sporangia are almost subhyaline, broadly ellipsoidal (Fig. 6.39).

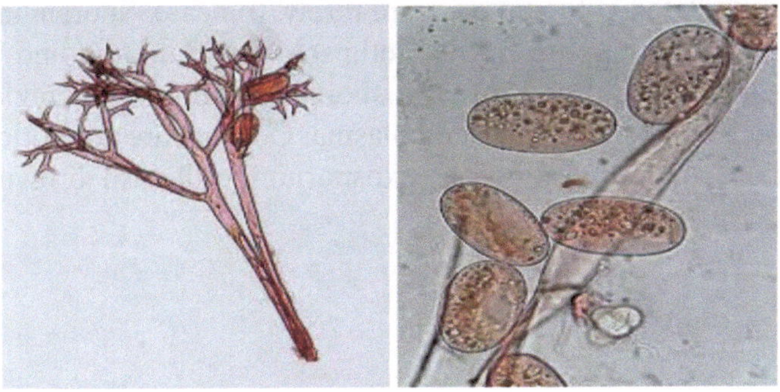

Fig. 6.39: Sporangiophore and sporangia of *Paraperonospora.*

6.2.5. *Bremia*

It has coarse intercellular mycelium. Haustoria are sac-shaped, many times they are present in each host cell. Sporangiophores are slender, more or less dichotomously branched, mostly at an acute angle. Sporangia usually germinates by single germ-tube or by zoospores. The tips of sporangiophores branches are enlarged into disks with sterigmata at the margin. Zoospores are rare. Parasitic effects on the plants of the family *Cichoriaceae* (Fig. 6.40).

Fig. 6.40: Sporangiophore and sporangia of *Bremia*.

6.2.6. *Bremiella*

Sporangiophores with determinate growth emerge singly or in groups through stomata and are hyaline, branched dichotomously in the upper part; the primary branches are of uniform thickness throughout. The ultimate branchlets are slightly inflated at the tip or, more rarely, truncate. Sporangia, borne singly at the branchlet tip, are hyaline, ellipsoidal with a pore and a conspicuous papilla at the distal end, often a short pedicellate. Sporangia germinates by zoospore or germinative naked plasma. Oospore are aplerotic, spherical, endosporium smooth, yellowish; exosporium yellowish brown, variously thickened (Fig. 6.41).

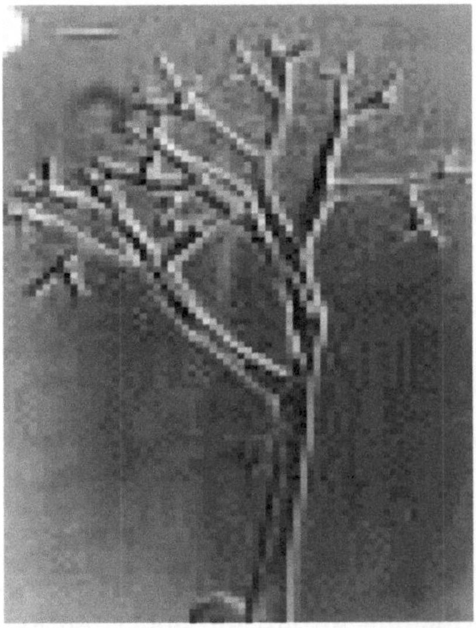

Fig. 6.41: Sporangiophore and sporangia of *Bremiella.*

6.2.7. Benua

The genus *Benua* is monotypic and possesses stout unbranched sporangiophores half-immersed in the stomata openings. The members of the new genus *Benua* exhibit several characteristics typical for general downy mildew with evanescent sporangia, including thin walled sporangiophores and sporangia and thick-walled oospores. Further differs from other GDM in having stout unbranched sporangiophores.

6.2.8. Sclerophthora

It develops thick-walled hyphae, and conidiophores are relatively short and develop from a basal cell. The unbranched or sympodial sporangiophores bear citriform or obpyriform sporangia. The sporangia are produced on the apex of the sterigmata and germinate by forming germ tubes. The number of zoospores produced in each sporangium varied from 4 to 32. These reproductive bodies are larger in size and different in structure. Oogonia are globose to sub-globose, thick-walled, hyaline, 20–48 × 10–29 μm, producing oospore-filling oogonia. Antheridia are hyaline, kidney-shaped, and paragynous. The mature oospore has a thick epispore, which fuses with the oogonial wall to form a thick brown covering over the oospore. The oospore is plerotic, which is characteristic of the genus *Sclerophthora*. The genus is parasitic on cereals and grasses (Fig. 6.42).

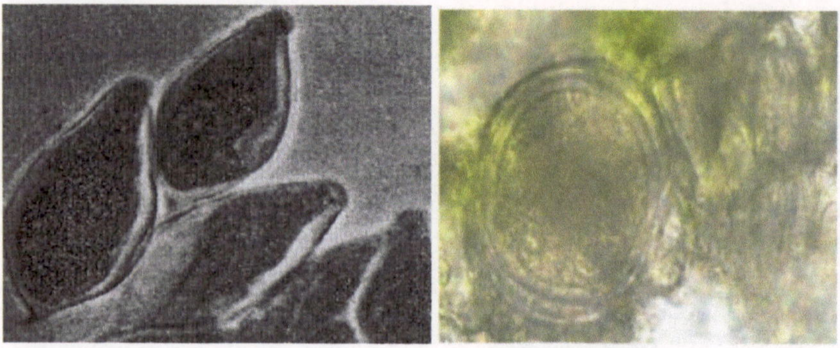

Fig. 6.42: Sporangium and oospore of *Sclerophthora.*

6.2.9. Peronosclerospora

The fungal pathogen has characteristics of both *Peronospora* and *Sclerospora.* Sporangiophores are erect, short, stout, widening toward the upper portion, and dichotomously branched 2–5 times at the apex, bearing sporangia on the sterigmata. Sporangia are hyaline, oval, spherical, subspherical, ovoid, cylindrical, elliptical, or oblong, with a round apex depending upon the species, thin-walled, non-papillate, and germinate by a germ tube-like *Peronospora.* Oospores are plerotic types, like *Sclerospora.* Oospores are spherical, light yellow, brown, or globular yellow, depending on the species. The genera are parasitic on maize, sugarcane, and sorghum (Fig. 6.43).

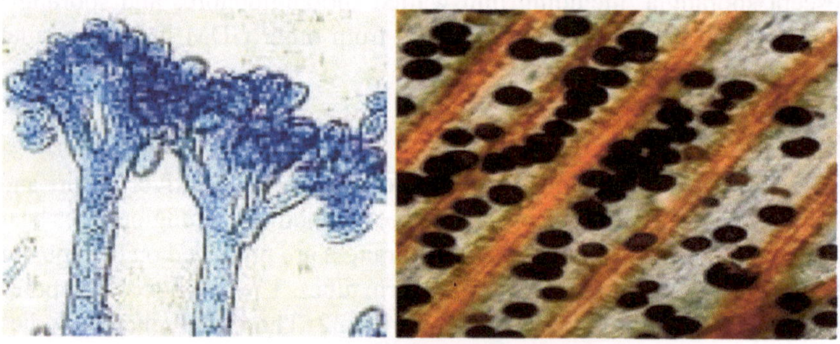

Fig. 6.43: Sporangiophore, sporangium and oospore of *Peronosclerospora.*

6.2.10. Hyaloperonospora

Sporangiophores are tree-like, down-hypophyllous, whitish, dense, and felt-like. Haustoria fills the host cell partly to almost completely. Conidiophores emerging through stomata, up to 30 in a fascicle, hyaline, somewhat stout, straight, 300–450 μm; trunk straight, 200–350 μm long, 10–13 μm wide. Ultimate branchlets are mostly in pairs, curved to sigmoid, 7–13 μm long, 1.5–2.2 μm wide at the base; the tip obtuse or subtruncate. Conidia are hyaline, broadly ellipsoidal to subglobose, 14.0–18.0 μm long, 11.3–15.5 μm wide, a length to width ratio of 1.11–1.29, the greatest width median, rarely sub- or supra-median, tip and base rounded; pedicel mostly present as a scar. Oogonia are irregular in shape, 36–57 μm in diameter; the wall are wrinkled, yellowish, and 2–5 μm thick. Oospores are plerotic to aplerotic, globose, 25–30 μm diameter, yellowish, 2–3.5 μm with a thick wall.

The genus *Hyaloperonospora* is the third largest genus of downy mildews, containing more than 100 species, which can infect economically important *Brassicaceae* crops such as horse-radish *(Armoracia rusticana)*, mustard greens *(Brassica juncea)*, rapeseed *(B. napus)*, tatsoi *(B. narinosa)*, cabbage *(B. oleracea var. capitata)*, cauliflower *(B. oleracea var. botrytis)*, Chinese cabbage *(B. rapa)*, arugula *(Eruca sativa)*, wasabi *(Eutrema japonicum)*, watercress *(Nasturtium officinale)*, and radish (*Raphanus sativus*) (Fig. 6.44).

Fig. 6.44: Sporangiophore, sporangium and oospore of *Hyaloperonospora.*

6.2.11. Basidiophora

Mycelium is coenocytic, intercellular with round knob like haustoria, 16–26 × 16–30 μm. *Basidiophora* produces small, unbranched sporangiophores with sporangia clustered at the apex. Sporangiophore are hyaline, straight, normally unbranched (abnormal branching reported in *Basidiophora entospora*), club-shaped, with a shaft 25 – 32.5 × 3 – 30 μm terminating in a swollen apex 10–30 μm wide, supporting 8–20 pedicellate sporangia, pedicels 4–15 × 1–4 μm. Sporangium formed singly on the pedicel, hyaline, ellipsoid, ovoid, limoniform or subspherical, 18–48 × 10–43 μm, wall smooth or verrucose, with an apical papilla 4–6 μm diameter. It produces zoospores (known only for *B. entospora*), reniform in shape, biflagellate, but small (4–5 × 3–4 μm). Oogonium formed intercellularly within the leaf tissue and oospores are spherical, with a smooth hyaline endospore and epispore and a golden yellow to brown, ridged or reticulate exospore. Parasitic on the plants of family Compositae (Fig. 6.45).

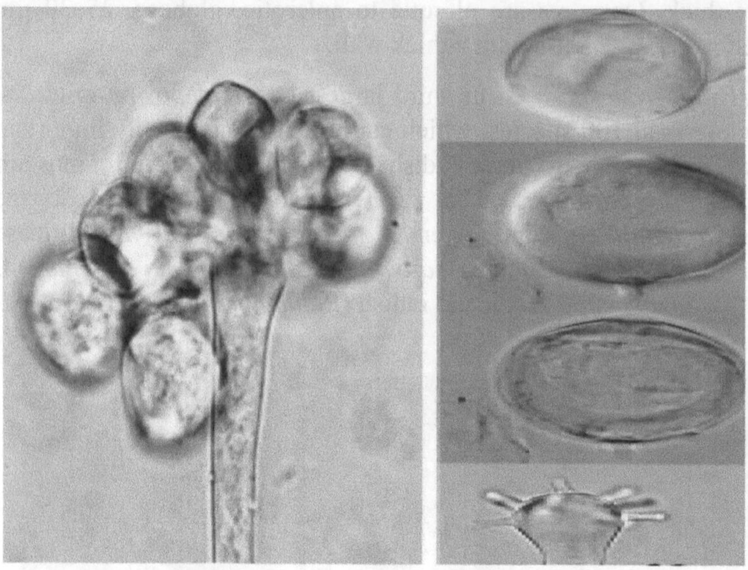

Fig. 6.45: Sporangiophore with pedicels and sporangium of *Basidiophora*.

6.2.12. Phytophthora

Mycelium is well developed, branched, coenocytic, hyaline, intercellular mycelium with haustoria. Thallus is heterothallic. Asexual reproduction is by zoospores produced in sporangia. Vesicle is not formed, and zoospores are differentiated in zoosporangium itself. Sporangium shapes ranged from spherical, ovoid, obpyriform, obturbinate, to ellipsoidal, but mostly papillate. Some were asymmetrical and others often had two divergent apices. Sporangia of *Phytophthora* causing downy mildew in lima bean averaging 39 × 22 μm and short pedicels, particularly diagnostic for *Phytophthora phaseoli*. Oospores are smooth, thick walled, round, dark brown and plerotic (Fig. 6.46).

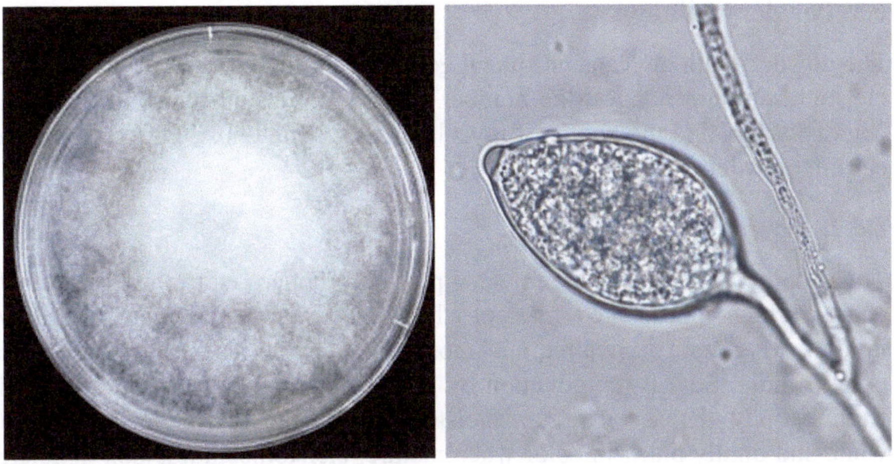

Fig. 6.46: Fungal culture of *Phytophthora* and sporangium.

Table 6.1: Differential characters of general *Phytophthora* and downy mildew *Phytophthora.*

Characteristics of Downy mildews	**Basal *Phytophthora* species**	***Phytophthora infestans***	***Sclerophthora***	***Poakatesthia***	***Viennotia***	**Dicot Downy mildews**
Sporongiophores generally emerge through stomata	–	(+)	+		+	+
Sporangia ripen simultaneously	–	–	–		+	+
Sporongiophore growth on host is determinate	–	–	+		–	+
Pathogens are obligate biotrophic	–	–	–	–?	+?	+
No intracellular mycelium except for haustoria	–	–	+?	–	+	+
Sporongiophore are lasting and well diffrentiated	–	(+)	–		+	+

+: yes; –: no; (): mostly; ?: not known with certainity.
doi:10.1371/journal.pone.0004790.t001.

6.3. Powdery Mildew

The powdery mildew-causing fungal genera are differentiated on the basis of cleistothecial appendages, the number of asci in cleistothecium, the type of conidial stage, and the nature of mycelium on infected host. Different genera causing powdery mildew are:

6.3.1. Erysiphe

The superficial mycelium on infected portion produces enormous number of conidia which appear like a mass of white powder, hence, the name powdery mildew. Mycelium is hyaline and mostly ectophytic. Asexual reproduction is by conidia. Sexual reproduction by gametangial contact producing sexual fruiting body the ascocarp i.e. cleistothecium. The cleistothecia are first white and finally black in color when mature. cleistothecium is non-ostiolate with flexible, flaccid, simple mycelioid appendages which resemble somatic hyphae. The cleistothecium contain several asci (Fig. 6.47).

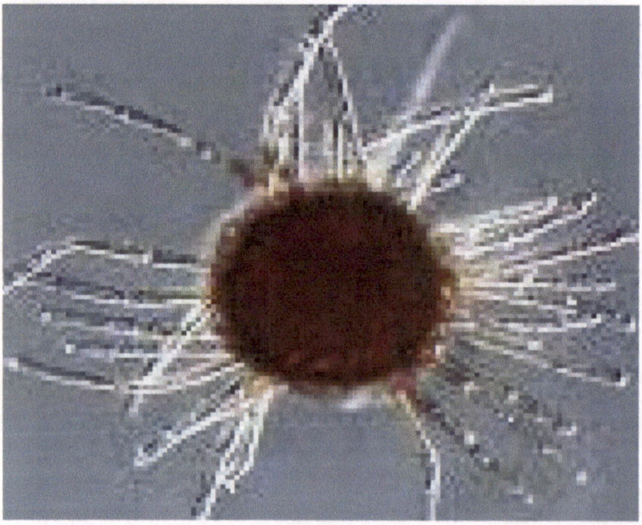

Fig. 6.47: Cleistothecial appendages of *Erysiphe.*

6.3.2. Microsphaera

Cleistothecium appendages are flattened and dichotomously branched. Cleistothecium contains several asci (Fig. 6.48).

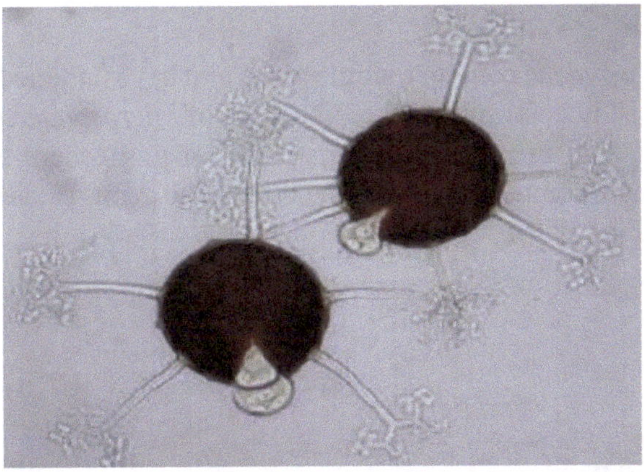

Fig. 6.48: Cleistothecial appendages of *Microsphaera.*

6.3.3. Phyllactinia

The cleistothecia of fungi in the genus *Phyllactinia* have two types of appendages, an equatorial group of radiating appendages that are unbranched and have bulbous bases (basal swelling) and a crown of repeatedly branched shorter appendages that secrete mucilage. The bases of the bulbous appendages are thick walled toward the base and thin walled toward the apex. As the cleistothecia dry, the appendages bend downward. This downward pressure of the appendage tips will lever the cleistothecia free from the mycelium. Cleistothecia contain multiple asci that each usually contain only two ascospores (Fig. 6.49).

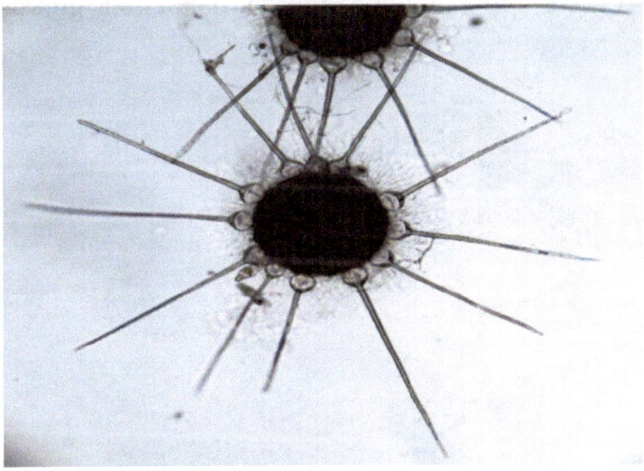

Fig. 6.49: Cleistothecal appendages of *Phyllactinia.*

6.3.4. Podosphaera

Cleistothecia of genera *Podosphaera* contain a single ascus and have flattened, dichotomously branched cleistothecial appendages like those of *Microsphaera*. Apart from the presence of only one ascus, the anamorph of this genera is quite distinct in having catenate conidia with fibrosin bodies. The genus causes powdery mildew on apple (Fig. 6.50).

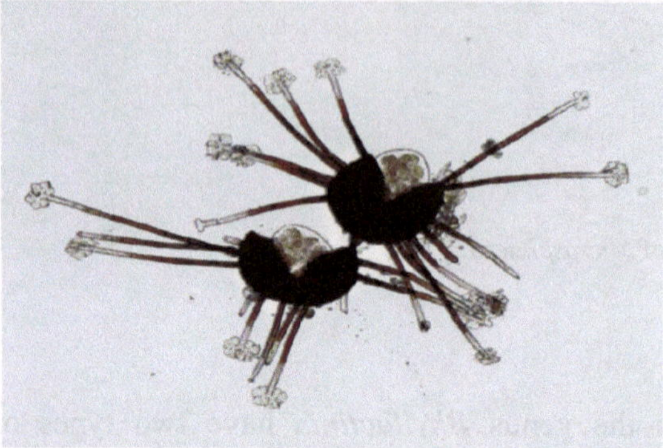

Fig. 6.50: Cleistothecial appendages of *Podosphaera.*

6.3.5. Sphaerotheca

Cleistothecia of this genus closely resemble *Erysiphe* cleistothecia, but it contains only one ascus per cleistothecium. The cleistothecial appendages are simple and non-branching.

The genus *Sphaerotheca* cause powdery mildew disease on roses, peach, gooseberry, hops, strawberry, currant, cantelope, squash, dandelion, gourds, and hackberry (Fig. 6.51).

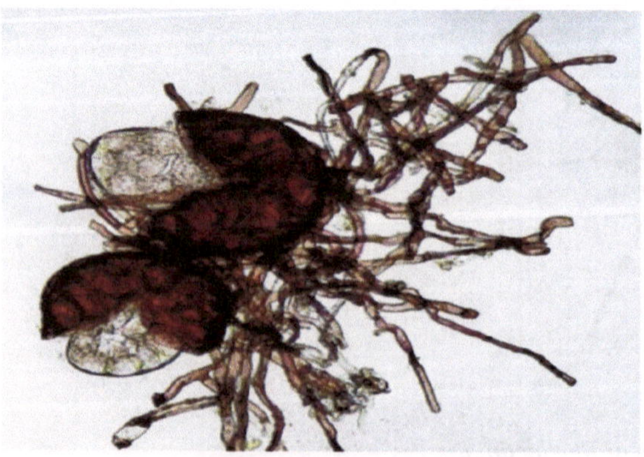

Fig. 6.51: Cleisthocial appendages of *Sphaerotheca.*

6.3.6. *Uncinula*

Uncinula genus is characterized by its dark chasmothecia and Cleistothecia which bear filamentous, hyaline appendages with hooked tips. One notorious species, *Uncinula necator* (syn. *Erysiphe necator*), attacks various species of grapes (*Vitis* spp.) and causes huge yield losses in European wine grapes. Other *Uncinula* species attack a wide variety of dicotyledonous plants (Fig. 6.52).

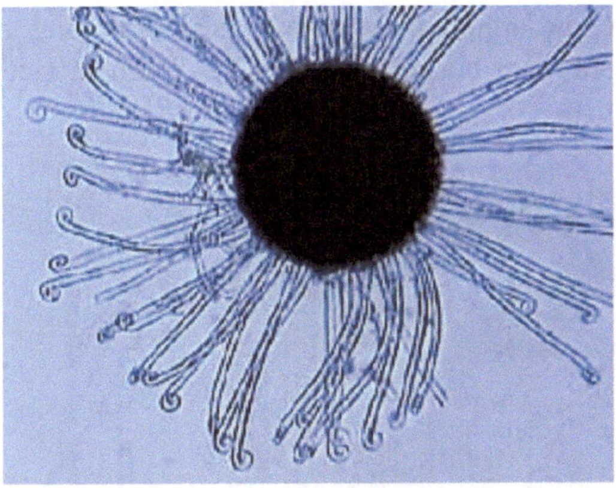

Fig. 6.52: Cleistohecial appendages of *Uncinula.*

6.3.7. *Leveillula*

The mycelium of the infected fungus is partly internal. Cleistothecial appendages are simple mycelioid. Cleistothecia are with several asci.

Anamorph stage of the pathogen produces conidia which are of two types i.e. having blunt tip and pointed tip. *Leveillula* sp. produce *Oidiopsis* as conidial stage. On the lesions white mycelium can be found with conidiophores bearing either lanceolate or rounded conidia (Fig. 6.53).

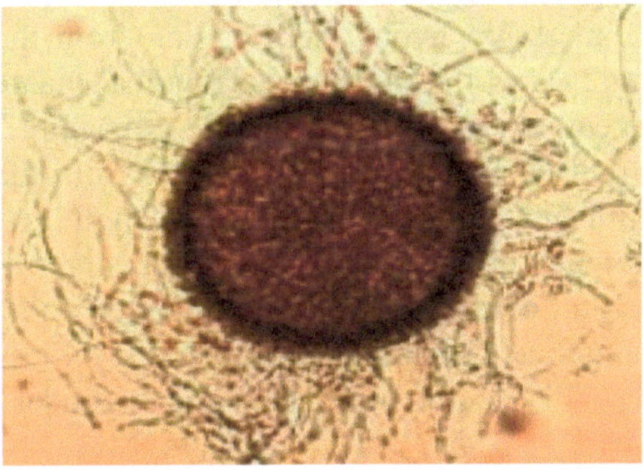

Fig. 6.53: Cleistothecial appendages of *Leveillula.*

6.3.8. *Blumeria*

The genus *Blumeria* differ from *Erysiphe* in its digitate haustoria and in details of conidia wall. *Blumeria* is considered to be phylogenetically distinct from *Erysiphe* as it hosts solely on true grasses of *Poaceae*.

Ascocarp is dark brown, globose with filamentous appendages, asci oblong, ascospore hyaline, ellipsoid, 20–30 × 10–13 μm in size.

Anamorph produces on hyaline conidiophores; catenate conidia of oblong to cylindrical shape, not including fibrosin bodies, 32–44 × 12–5 μm in size. Haustoria are palmate (Fig. 6.54).

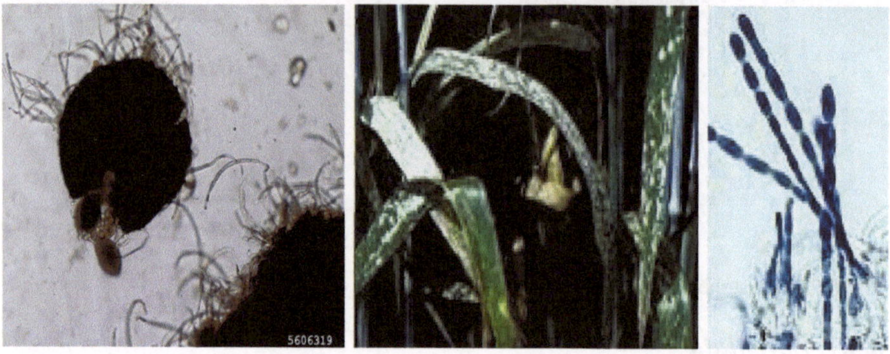

Fig. 6.54: Cleistothecial appendages of *Blumeria*, infection on wheat and anamorph oblong conidia.

6.3.9. *Golovinomyces*

Golovinomyces is characterized by having chasmothecia and cleistothecia with mycelioid appendages and several, mostly two-spored asci. An asexual morph with catenescent conidia that lack fibrosin bodies and mostly nipple-shaped appressoria. Conidiophore single on hyphal cell, arising from the upper part of mother cell, producing 2–4 swollen cells in chain with a sinuate outline, 120–200 μm long, and foot cell straight, 65–120 × 9–12 μm, simple, with a basal septum at the branching point or slightly displaced from the mycelium. Conidia are oval to ellipsoid, 30–42 × 20–26 μm (length and breadth ratio 1.3–1.9), lacking distinct fibrosin bodies, producing germ-tubes on the perihilar position, with reticulate wrinkling of the outer walls. First formed conidia are apically conical, basally round to more or less subtruncate, and generally shorter (30–36 μm long) than the secondary conidia (Fig. 6.55).

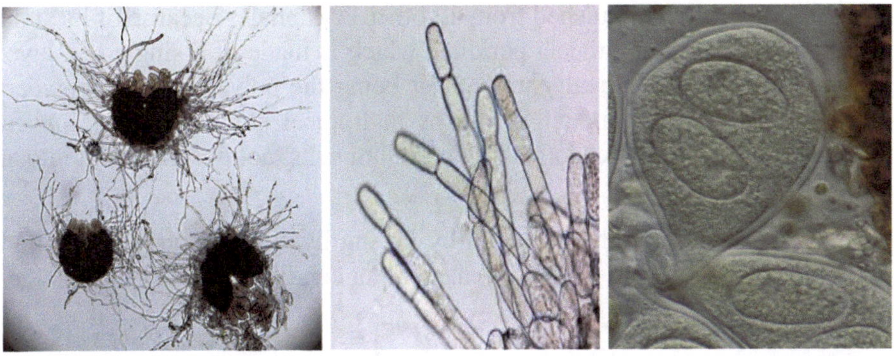

Fig. 6.55: Cleistothecial appendages of *Golovinomyces*, two spored asci and anamorph conidia.

6.3.10. Sawadaea

Cleistothecia with several asci, cleistothecial appendages stiff, radiating with dichotomously or trichotomously branched, curved tips. Conidia with fibrosin bodies. The genus is almost confined to maples (Fig. 6.56).

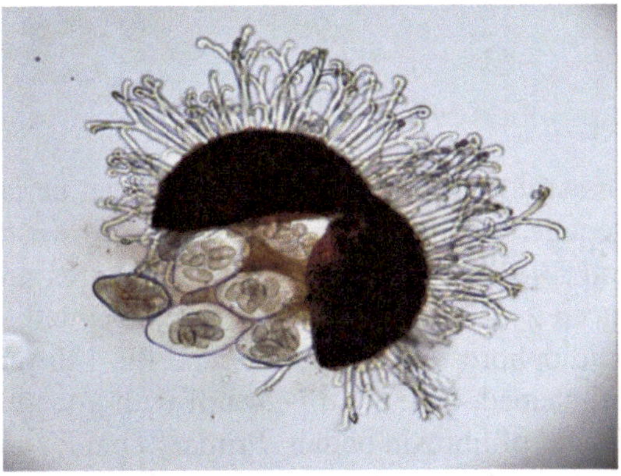

Fig. 6.56: Cleistothecial appendages of *Sawadaea.*

6.3.11. Brasiliomyces

The fungus forms scattered colonies of white, superficial mycelium on the host leaves. Small globose, hyaline, enclosed ascocarps are found on the surface of mycelium. Each ascocarp contains three saccate asci; each asci typically contains five oval, hyaline, and one-celled ascospore, but 3–6 ascospores are also observed (type sp. *B. malachrae*).

Brasiliomyces is distinguished from all other Erysiphales because of its "thin, semitransparent chasmothecia peridia," which is made of a single cell layer. In addition, it is ecologically unique for being the only species of powdery mildew that forms abundant ascocarps in tropical climates. *Brasiliomyces* have a single layer of cleistothecium peridium cells, not differentiated into an inner and outer layer.

The asexual morph is characterized by having hyaline ectophytic mycelium with lobed hyphal appressoria and catenescent conidia (Fig. 6.57).

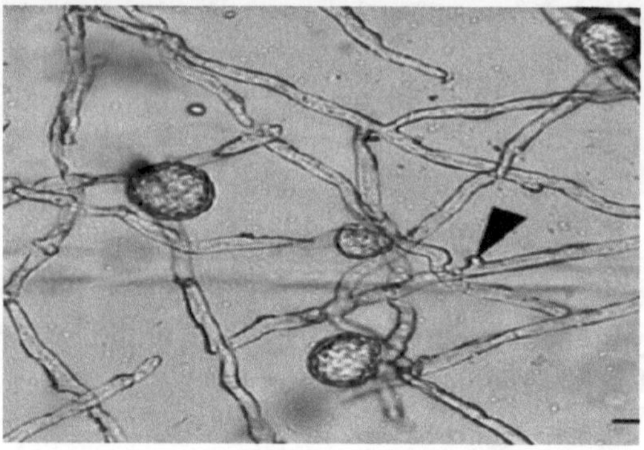

Fig. 6.57: Cleistothecium of *Brasiliomyces.*

6.3.12. Cystotheca

This powdery mildew genus is mostly present on Oak trees. Special aerial hyphae are falcate to curved, aseptate, at first hyaline, later deep brown to purplish brown, thick-walled, and 80–140 × 6–15 µm. Conidiophores are 175–245 × 10–12 µm, and produced 2–4 immature conidia in chains with a sinuate outline. Foot-cells of conidiophores are cylindrical and 80–120 µm long. Conidia are lemon- to barrel-shaped, 26–35 × 17–24 µm with a length and width ratio of 1.2–1.6, and devoid of fibrosin bodies. Primary conidia are apically conical and basally subtruncate. Germ tubes produce in the lateral to perihilar position of conidia are long and slender (3 µm wide). Chasmothecia are scattered or gregarious, partly embedded with special aerial hyphae, dark brown, spherical, 55–70 µm in diameter, and contain a single ascus. Chasmothecial peridia consisted of two layers. Exoperidia (outer layer) are composed of dark brown, polygonal cells 10–20 µm wide. Endoperidia (inner layer) consist of hyaline, polygonal cells 10–15 µm wide. Appendages are basally attached, mycelioid, rare or few, and pale brown to rusty brown. Asci are short stalked, 55–72 × 35–46 µm, have a terminal oculus 10–20 µm wide, and contained eight ascospores. Ascospores are oblong-elliptical, 22–27 × 10–12.5 µm, subhyaline, contains one or two oil drops (Fig. 6.58).

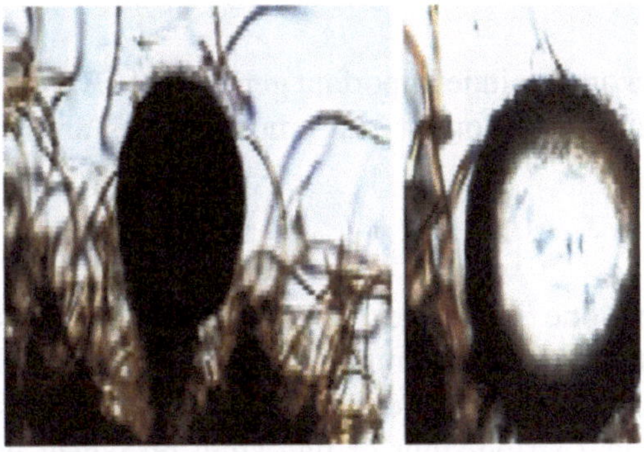

Fig. 6.58: Cleistothecium of *Cystotheca.*

6.4. Gray Mildew

On older leaves, small, light green to yellowish, angular spots that are limited by the veins appear on the upper surface. Whitish grey powdery growth just beneath the spots on lower side.

6.4.1. Ramularia

The fungus produces endophytic, septate mycelium. Conidiophores are short, hyaline and branched at the base. Conidia are borne singly or in chains at the tips of conidiophores. The conidia are hyaline, irregularly oblong with pointed ends, sometimes rounded to flattened ends, unicellular or 1–3 septate. The perfect stage of the fungus produces perithecia containing many asci. The ascospores are hyaline and usually two-celled (Fig. 6.59).

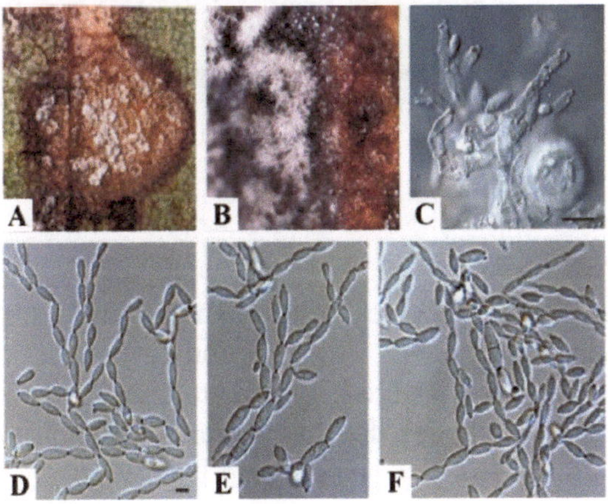

Fig. 6.59: Perithecia (A–C) and conidia (D–F) of *Ramularia.*

6.5. Leaf Smut

Leaf smut fungi are *Basidiomycetes* and include important genera like *Ustilago* and *Urocystis*. The most economically important leaf smut diseases are on cereals and grasses.

6.5.1. Urocystis

Urocystis teliospore consists of one or more melanized fertile cells surrounded by several sterile cells. This type of compound teliospore is called a spore ball. There are about 140 species in this genus. An important species is *Urocystis tritici* which causes leaf stripe-smut or flag smut on wheat in warm climates. The fertile cell of a spore ball of *Urocystis* germinates by producing an aseptate promycelium which gives rise to about four primary sporidia. Teliospores of *U. tritici* survive in the soil and attached to seeds (Fig. 6.60).

Fig. 6.60: Spore ball of *Urocystis* and infection on leaf.

6.5.2. Ustilago

Ustilago genus of smut fungi is parasitic on cereal grasses and several other plants species. Teleospores are single and small. The infection develops into sori and the black dusty teliospores are covered by a membrane of host origin. This thin membrane breaks to expose the dusty spores at maturity (Fig. 6.61).

Fig. 6.61: *Ustilago* spores and smut sori on infected leaves.

6.6. Leaf Galls and Blisters

Important fungal genera causing leaf galls are *Taphrina* and *Exobasidium.*

6.6.1. Taphrina

Taphrina genus is an ascomycetes fungus that causes leaf gall, catkin curl diseases, and witch's brooms in certain flowering plants. One of the more commonly observed species causes peach leaf curl. *Taphrina* typically grow as yeasts during one phase of their life cycles, then infect plant tissues in which typical hyphae are formed, and they form a naked layer of asci on the deformed, often brightly pigmented surfaces of their hosts. No discrete fruit body is formed outside of the gall-like or blister-like tissues of the hosts. The asci form a layer lacking paraphyses, and they lack croziers. The ascospores frequently bud into multiplc yeast cells within the asci (Fig. 6.62).

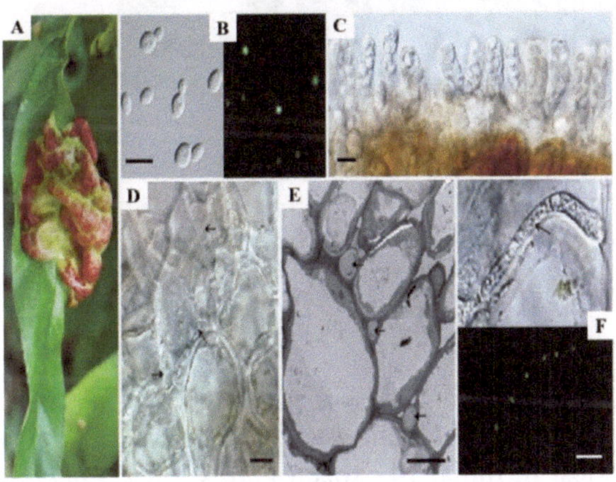

Fig. 6.62: *Taphrina* leaf gall and open ascus of the fungi.

6.6.2. Exobasidium

Exobasidium is a genus of fungi in the family *Exobasidiaceae*. Species of *Exobasidium* are parasitic on angiosperms; preferred hosts are temperate *Ericales*. Infected host tissues on leaves, succulent shoots, flowers, and fruits form misshapen galls or blisters, which often are discolored, white, or pink. Basidia develop in a dense layer on the outer surfaces of the infected host tissues, often protruding from stomata or from between epidermal cells. It possesses a haploid saprotrophic yeast-like phase and a dikaryotic biotrophic phase. They do not produce teliospores, but basidia are formed directly on the surface of the infected host. The mycelium is usually intercellular, with haustoria often present. A whitish superficial layer of basidia is eventually produced on the surface of infected host tissue, and basidiospores are produced on the basidial sterigmata. Basidiospores are initially aseptate, but in many species they become septate after their discharge, and both cells can germinate by emitting germ tubes (Fig. 6.63).

Fig. 6.63: Basidium, basidiospore and blisters on tea leaves.

6.7. Rust

The rust genera are differentiated on the basis of uredospore and teliospore, and their shapes.

6.7.1. Puccinia

All species in the *Puccinia* genus are obligate plant pathogens and cause rust disease. It has 1800 species. In India, 147 species are known. *Puccinia* forms five types of spores: urediniospores/uredospore, teleutospore/teliospore, basidiospores, pycnidiospore, aeciospores or spermagonium (in wheat rust, the first three spores are formed on the wheat host, while the last two stages are formed on the barberry). However, in the infected host disease samples, two types of spores are usually noticed, i.e. uredospores and teliospores. Uredospores are stalked, unicellular, rounded, binucleated spores produced during the early symptoms of the development of rust sori. During the uredospore development stage, the mycelium produces a number of reddish-brown pustules or streaks on the upper surface of the leaf known as uredosorus, consisting of a number of uredospores. The epidermis of the host plant bursts, and uredospores are liberated, carried by the wind and attack fresh plants. The uredospore enters through the stoma and produce dikaryotic mycelium, which produces uredospores within 6–10 days, so are called repeating spore.

During the later stage, the dikaryotic mycelium of *Puccinia*, which gives rise to uredospores, now produces teliospores. The teliospores are produced in the teliosorus. Teliospores are bi-celled, stalked, and binucleate. The teliospores are liberated by breaking the epidermis and they do not attack the plant. The two nuclei in a cell of teliospore fuse to form a diploid nucleus. They remain dormant in the soil until the beginning of the next crop season. During the germination of teliospores, the teliospore produces one germ tube from each cell. The germ tube is known as promycelium or epibasidium. The diploid nucleus in a cell of the teliospore moves into the promycelium and divides meiotically to form four haploid nuclei: two are of (+) strain and two are of (−) strain to produce basidiospores to infect alternate hosts and complete other stages in the life cycle of the pathogen (Fig. 6.64).

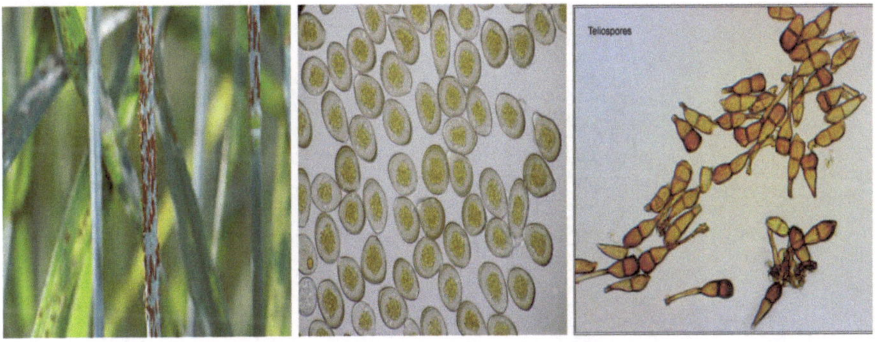

Fig. 6.64: *Puccinia* rust pustule on wheat plant, one celled uredospore and two celled teliospore.

6.7.2. Uromyces

In Uromyces the teliospores are single celled with a thick apex (papillum) and stalked. The stalks are fragile and short. The uredospores are yellowish brown, elliptical to oval with spiny surface (Fig. 6.65).

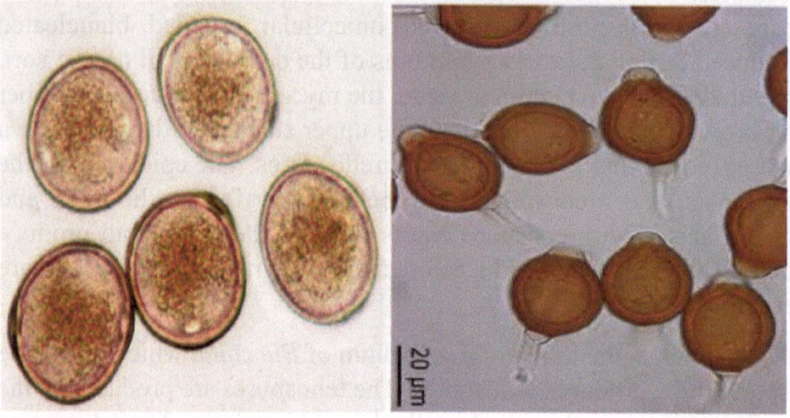

Fig. 6.65: Uredospore and Teliospore of *Uromyces.*

6.7.3. Gymnosporangium

Gymnosporangium telia are golden brown, 10–22 mm long and 1–2 mm wide and appear as hornlike projections from round galls on the infected host. The telia consist primarily of two-celled, cylindric fusoid teliospores. Teliospores are thick-walled, yellow to golden in color and have two pores near the septum. It produces dikaryotic mycelium which can survive for many years in the principal host, and repeated production of telia is associated with a spindle-shaped swelling or canker of the infected trunk or twig (Fig. 6.66).

Fig. 6.66: Telia of *Gymnosporium* on host and teliospores.

6.7.4. *Cronartium*

Cronartium genus is heteroecious rusts with two alternating hosts, typically a pine and a flowering plant, and having five spore stages. *Cronartium* produces one celled teliospores in long columns projecting out of the telia. The teliospores are attached to each other but not to their host tissue, and they are unstalked and sessile (Fig. 6.67).

Fig. 6.67: Telia of *Cronartium* and unstalked teliospore.

6.7.5. *Hemileia*

Uredospores formed in substomatal chamber are reniform, hump backed resembling orange segments, with echinulate upper and smooth lower surface. Teliospores arise through stomata germinate immediately after they are formed. Teliospores are produced in small numbers, they are turnip shaped, one celled, stalked, smooth walled on erumpent, club shaped stalks. It forms a mixed sorus consisting of both uredo and teliospores. Pycnia and aecia are not known (Fig. 6.68).

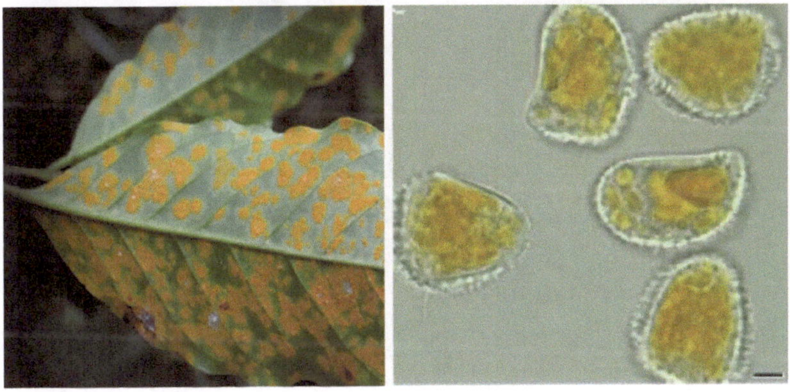

Fig. 6.68: *Hemileia* infection on coffee leaf and uredospores.

6.7.6. *Melampsora*

Melampsora teliospores are single-celled, sessile, or with a stalk while uredospores are single-cell sessile, and both are formed in a palaside layer. The telia are erumpent and covered by the host epidermis. The pathogen is autoecious, meaning it completes its life cycle on a single plant (Fig. 6.69).

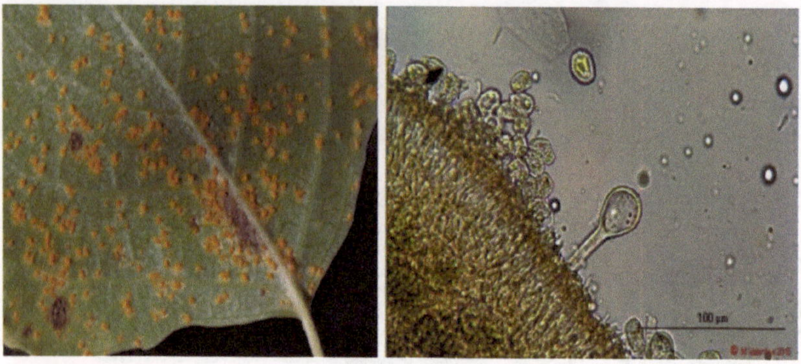

Fig. 6.69: *Melampsora* rust on poplar and uredospores in rust sori.

6.7.7. *Phragmidium*

Phragmidium genus typically infects plant species in the family Rosaceae. It is characterized by having stalked teliospores borne on telia, each having a row of four or more cells. The teliospores contain mostly 4 (*Phragmidium violaceum*) or 6–8 (*Phragmidium mucronatum*) cells with very dark and rough walls, and are borne on a long stalk (pedicel) which becomes easily detached from the lesions (Fig. 6.70).

Fig. 6.70: *Phragmidium* on rose leaf with uredospore and teliospore.

6.7.8. Phakopsora

Phakopsora infection is generally hypophyllus, growing on the undersurface of leaves. Uredinia on *Vitis hypophyllous,* minute, scattered or aggregated in small groups, subepidermal, becoming erumpent, and surrounded by abundant paraphyses. Paraphyses cylindrical to weakly incurved, evenly thin-walled or dorsally thick-walled (1.5–4.0 µm), 30–75 µm high. Uredospores are obovoid, obovoid-ellipsoid, or oblong-ellipsoid, 15–29 × 10–18 µm. Spore wall is approximately 1.5 µm thick and uniformly echinulate. Six or rarely four germ pores, equatorial. Telia hypophyllous, crustose, brown to blackish-brown, often confluent, sub-epidermal, applanate. Teliospores are regularly arranged in 3–5 layers, oblong to oblong-ellipsoid, 13–32 × 7–13 µm. The spore wall is evenly thin, slightly thickened, and brownish in the spores of the uppermost layer. Basidiospores reniform, 8–11 × 5–8 µm (Fig. 6.71).

Fig. 6.71: *Phakopsora* infection on grape leaf, rust uredia and urediospores.

6.7.9. Coleosporium

Coleosporium infection is found in Plumeria species. The spermagonial and aecial stages are unknown. Urediospores are catenulate, verrucose, with scattered germ pores. The urediospore ornamentation consists of distinctive, annulate, bluntly capitate tubercles. Light microscopy of telia indicates that teliospores are smooth, gelatinous, and one-celled, quickly appearing four-celled with division of the protoplast into an internal four-celled basidium (Fig. 6.72).

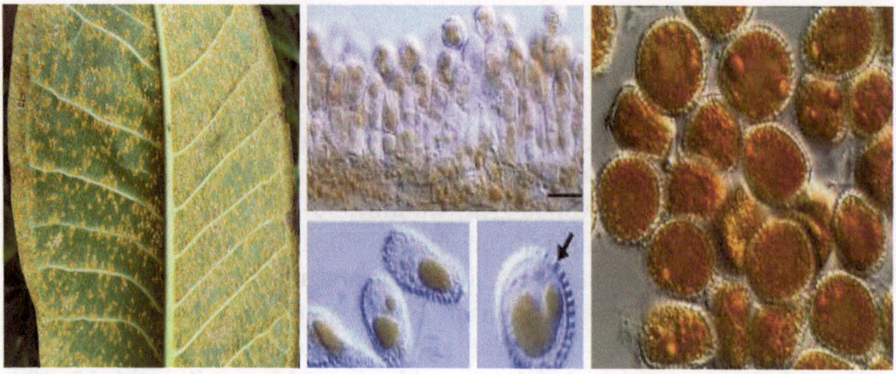

Fig. 6.72: *Coleosporium* sori on plumeria with ornamentation on urediospores and teliospores.

6.7.10. Chrysomyxa

The genus contains about 23 species is widespread in the Northern Hemisphere. Rust fungi in the genus *Chrysomyxa* occur in boreal forests of the northern hemisphere on Pinaceae (mostly *Picea*), and most species alternate to angiosperm hosts in the Ericaceae. Diseased cones open prematurely and orange–yellow aeciospores are sometimes produced in such quantity as to color the forest floor and lake surfaces. Up to half of a cone crop can be affected, and most of the affected cones do not produce viable seed. They sporulate aecia and blackish spermogonia. Aeciospores had a distinct and narrow longitudinal groove on one side and tapered warts on the surface. The peridial cell wall is thick (7.18.7 µm), with a coarse and striate surface. Aeciospores are 13.130–14.843 µm (Fig. 6.73).

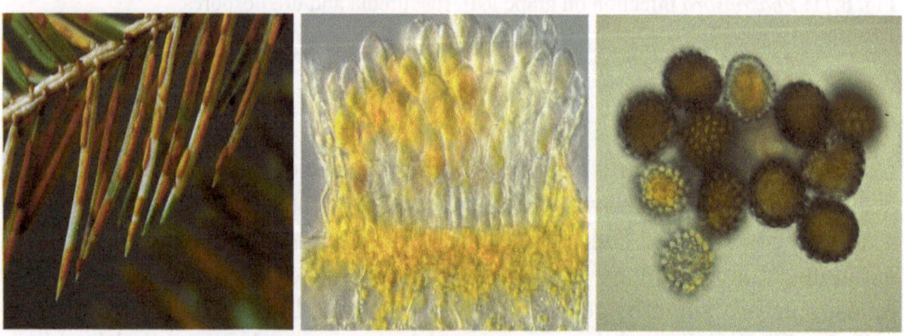

Fig. 6.73: *Chrysomyxa* rust infection on pine needles and the rust spores.

6.7.11. Physopella

Infection spots epiphyllous, appearing as black, shining crusts resembling tar spots. Telia are sub-epidermal, non-erumpent, consisting of laterally and terminally united teliospores. Telial sorus spreading, 1.9–2 mm. deep below the epidermis. Teliospores catenulate in sessile chains of 3–5 spores, one celled, measuring 14.4–27 × 9–21.6 μm, germ pore obscure, wall thick 2–3.6 μm, pigmented brown. Terminal teliospores are often curved or beaked, with a wall 3.6–7 μm in thickness, germination without dormancy, and basidium external. Uredospores develop just below the epidermis. Most of the intercellular hyphae away from the sporulating zone are highly vacuolated. Uredospore is binucleate. The non-erumpent telia and the catenulate arrangement of the teliospores are characteristic of *Physopella* (Fig. 6.74).

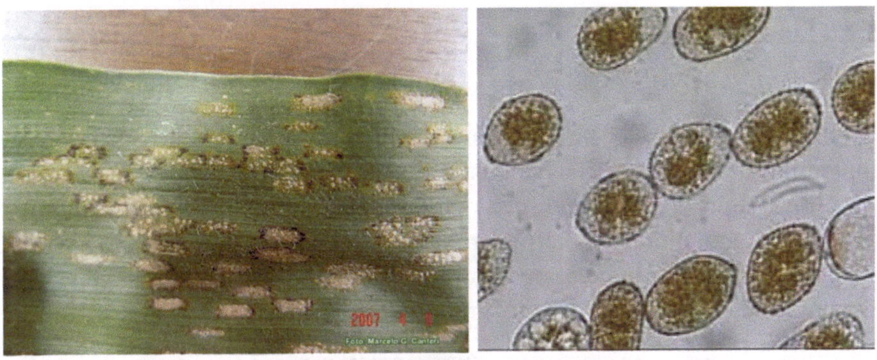

Fig. 6.74: *Physopella* rust infection on maize and rust uredospore.

6.8. White Rust

6.8.1. *Albugo*

Albugo is a genus of plant-parasitic oomycetes which causes white rust or white blister diseases in above-ground plant tissues in cruciferous crops. The rust sori of Albugo produces sporangia on sporangiophores wherein biflagellate zoospores of about eight per sporangium is formed. Sporangiophores are 30–45 × 15–18 μm, non-septate, hyaline. Sporangia in basipetal chains (oldest at top), globose to oval, 12–18 μm diam., hyaline, thin-walled. Oospores rarely found in leaves, usually in stems, globose, brownish, 30–55 μm diameter, wall thick, tuberculate with low blunt ridges that are often confluent and irregularly branched. The *Albugo* reproduce by producing both sexual spores (i.e. oospores) and asexual spores (i.e. sporangia) (Fig. 6.75).

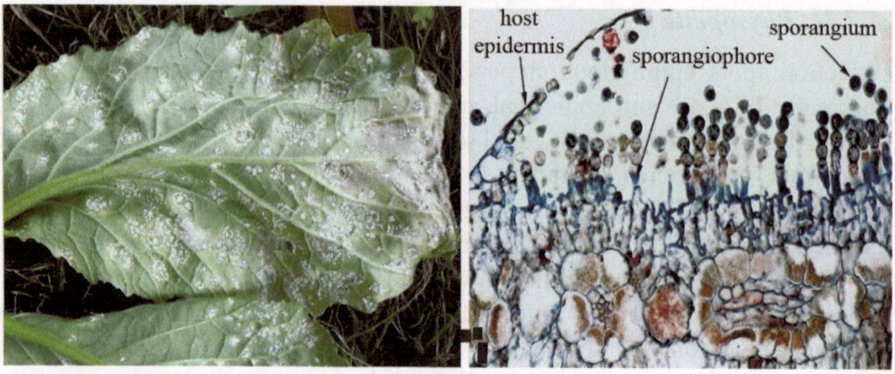

Fig. 6.75: White rust on crucifers and rust pustule with sporangium.

6.9. Cane Red-Rot

6.9.1. Colletotrichum

The fungus produces thin, hyaline, septate, profusely branched hyphae containing oil-droplets. The fungus produces black, minute velvety acervuli with long, rigid bristle-like, septate pointed setae on the surface of rind, leaf midrib and sometimes in the pith region. Conidiophores are closely packed inside the acervulus, which are short, hyaline and single celled. The conidia are single-celled, hyaline, falcate, granular, and guttulate (Fig. 6.76).

Fig. 6.76: Sugarcane red rot and conidia in acervulus.

Five pathotypes of the fungus have been reported. The perfect stage of the fungus produces large number of globose and dark brown to black perithecia with a papillate ostiole. Asci are clavate, unitunicate and eight spored. Large numbers of hyaline, septate, filiform paraphyses is also present among asci. In culture, most of isolates produce cottony, fluffy, or suppressed colonies.

7

Identification of Fungi Causing Stem and Twig Diseases

7.1. Die-Back

7.1.1. Colletotrichum

The pathogen forms a large number of acervuli on the infected parts. The conidiophores are slightly curved, short, and club-shaped. The conidia are hyaline and falcate, borne singly on the conidiophores. Numerous black-colored and thick-walled setae are also produced in acervulus. In culture, most isolates produce cottony, fluffy, or suppressed colonies (Fig. 7.1).

Fig. 7.1: Die-back on orange plant and acervulus of the pathogen.

7.1.2. Diplodia

Diplodia has uni- or multi-locular conidiomata lined with conidiogenous cells that form hyaline, aseptate, and thick-walled conidia at their tips. Typically, the conidia remain hyaline for a long time before they become brown and septate.

Diplodia is a genus of anamorphic fungi in the family *Botryosphaeriaceae*. *Diplodia* produces dark brown to black pycnidia after 14–20 days of incubation with hyaline conidiogenous cells that are cylindrical, thin-walled, and smooth and produce a single conidium at the tip (Fig. 7.2).

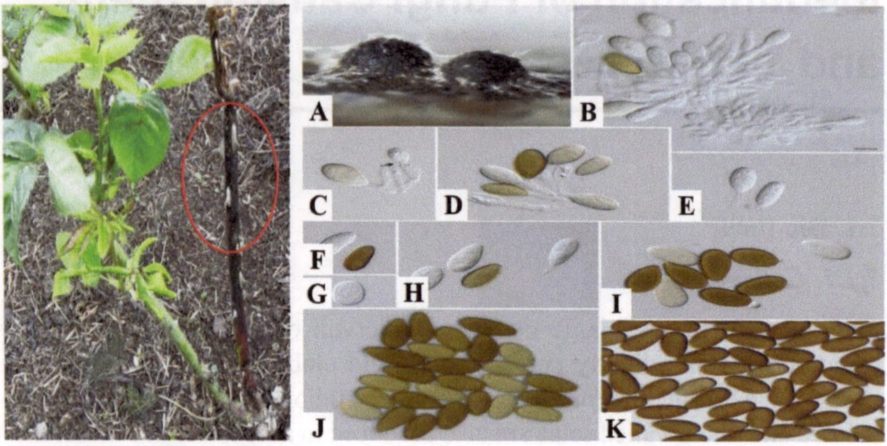

Fig. 7.2: *Diplodia* die-back on rose plant, pycnidia (A) and *diplodia* spores (B to K).

7.1.3. Phomopsis

The mycelium of the pathogen is septate and hyaline, becoming dark with age. Pycnidia are submerged and later become erumpent with a prominent ostiole. Conidia are produced on simple to branched conidiophores and are of two types; alphaconidia, which are subcylindrical and beta conidia, which are filiform and curved. The perfect stage produces perithecia, in which asci with eight hyaline, bi-celled, ellipsoid fusoid ascospores are produced, which are usually constricted at the septum (Fig. 7.3).

Fig. 7.3: *Phomopsis* twig blight with pycnidia and pycniospores.

7.1.4. Eutypa

Eutypa dieback is caused by the fungus *Eutypa lata* which survives in infected trunks for long periods of time. The fungus produces reproductive structures (perithecia) on the surface of infected wood. Spores (ascospores) are produced in these structures and are discharged into the air (Fig. 7.4).

Fig. 7.4: *Eutypa* infection on grapevine and ascospores produced in perithecia.

7.1.5. Lasiodiplodia (Botryodiplodia)

Lasiodiplodia genus belongs to the family *Botryosphaeriaceae*. The mycelial colony of *Lasiodiplodia theobromae* produced very thin, uniform, thread-like growth in all media. The mycelial growth is blackish on potato dextrose agar and Czepek's Dox agar, while it is whitish on peptone salt agar and oatmeal agar media.

Pycnidium is carbonaceous, dark brown or black, clustered, and non-ostiolate. Conidiophores are simple and short and show distinct annellations and conidia are dark brown, 2-celled, ovoid with characteristic striations at maturity (Fig. 7.5).

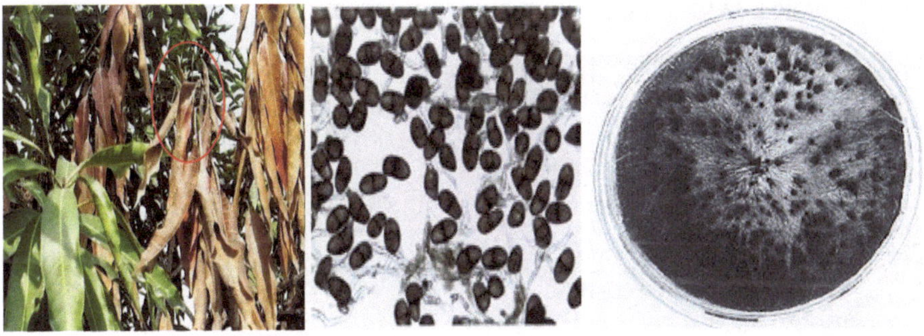

Fig. 7.5: *Lasiodiplodia* dieback of mango, its conidia, and pycnidium on growth media.

7.1.6. *Neofusicoccum*

Neofusicoccum genus belongs to the family *Botryosphaeriaceae*. *Neofusicoccum parvum* is one of the most aggressive causal agents of *Botryosphaeria* dieback. The fungal isolates are fast-growing and form many aerial mycelia, which are initially white but turn gray to black over time. The reverse of the colonies is white, which also became grayish to black with aging. The conidial size ranged from 17.5 to 25.0 µm (avg. 20.8) × 7.5–10.0 µm (avg. 7.9). Pycnidia are produced in PDA, whose size varies from 370.6 to 908.9 × 234.5–506.4 µm (Fig. 7.6).

Fig. 7.6: *Neofusicoccum* die-back on avocado and the conidia.

7.1.7. *Phaeoacremonium*

Phaeoacremonium genus is associated with wilt and decline diseases of woody hosts and human infections. Cultures are usually slow-growing, suede-like with radial furrows, initially whitish gray becoming olivaceous gray with age. Hyphae hyaline, later becoming brown and some becoming rough-walled. Phialides are brown, thick-walled, slender, acular to cylindrical, slightly tapering toward the tip, often proliferating, with small, funnel-shaped collarettes. Conidia, often in balls, are hyaline, thin-walled, cylindrical, or sausage-shaped (Fig. 7.7).

Fig. 7.7: *Phaeoacremonium* die-back on oak, its conidia and fungal mycelial growth on media.

7.1.8. Paraconiothyrium

Conidiomata 250–350 µm high × 200–300 µm diameter, pycnidia solitary, immersed, globose, unilocular, centrally ostiolate. Conidiomata wall 15–25 µm wide, 4–5 layered, composed of outer 3–4 layers brown and inner 1–2 layers hyaline, thin-walled cells of textura angularis. Conidiophores are reduced to conidiogenous cells, arising from the base and sides of the conidioma. Conidiogenous cells are 3.5–6.5 × 3–4 µm in size, phialidic, with a minute collarette, globose to coliform, lining the inner wall layer of the pycnidium, hyaline, and smooth. Conidia are 4.5–6.5 × 3–4.5 µm in size, solitary, hyaline when attached to conidiogenous cells, becoming dark brown, globose to obovate, one-celled, thick, and smooth-walled. Fungal colonies are colorless to buff margin, immersed mycelium, a dense moderately high mat of woolly-floccose, white to grayish, in the center weakly hazel aerial mycelium, conidiomata observed (Fig. 7.8).

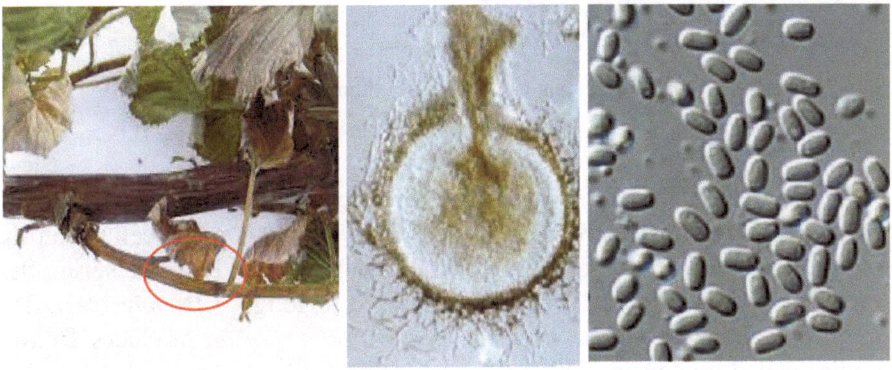

Fig. 7.8: *Paraconiothyrium* twig blight of raspberry, pycnidia and conidia.

7.1.9. Ceratobasidium

Ceratobasidium genus belongs to the order *Cantharellales*. Basidiocarps (fruiting bodies) are effused, and the genus is sometimes grouped among the corticioid fungi. *Rhizoctonia*-like anamorphs of *Ceratobasidium* species are placed in the genus *Ceratorhiza*. Basidiocarps are effused, thin, and often inconspicuous; smooth; waxy to dry and web-like; whitish to pale gray. Microscopically, they have comparatively wide hyphae without clamp connections and basidia that are spherical to cuboid or broadly club-shaped. Basidia bear 2 to 4 sterigmata, which are comparatively large. Basidiospores are globose to cylindrical (elongated and worm-like in the type species), smooth, and colorless. They frequently produce secondary spores and germinate in hyphal tubes. *Ceratorhiza* anamorphs produce hyphae (sometimes swollen) and occasionally sclerotia (small propagules composed of thick-walled hyphae) (Fig. 7.9).

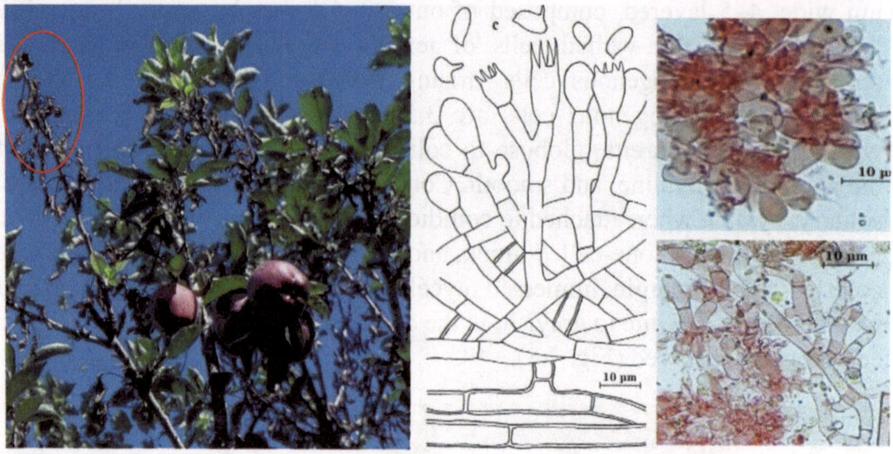

Fig. 7.9: *Ceratobasidium* twig blight on apple and basidiocarp.

7.1.10. Sclerotinia

A key characteristic of this pathogen is its ability to produce black resting structures known as sclerotia and white fuzzy growths of mycelium on the plant it infects. These sclerotia give rise to a fruiting body in the spring that produces spores in a sac, hence the name "sac fungi" (*Ascomycetes)*. The mycelium then covers the necrotic area. *S. sclerotiorum* produces fruiting bodies called apothecia, which are small, thin stalks ending with a cup-like structure about 3–6 mm in diameter. The cup of the apothecium is lined with asci, in which the ascospores are contained (Fig. 7.10).

Fig. 7.10: *Sclerotinia* stems rot in soybean, sclerotia, and apothecia with asci.

7.1.11. Hymenoscyphus

Hymenoscyphus genus belongs to the family *Helotiaceae*. It produces a sexual fruiting body i.e. white stalked apothecia (mostly up to 3 mm in diameter, rarely 8 mm), on the leaf litter of the previous year. The hymenium is composed of cylindrical paraphyses (1.8–2.4 µm thick, enlarged to 3 µm at the apex) and cylindrical clavate asci (80–107 × 8–10 µm), showing a positive iodine reaction and forming hyaline single-celled ascospores (13–21 × 3.5–5.0 µm). The fungal colony is effuse, cottony, dull white to fulvous brown (dull reddish-yellow/orange or brown-yellow), with some patches becoming gray to dark gray, 9–28 mm in diameter. Vegetative hyphae are subhyaline to olivaceous brown, 1.2–3.0 µm wide with rare swellings up to 4.2 µm, thin-walled, smooth, and septate with septa 5–21 µm apart. Conidia are unicellular, hyaline to subhyaline, filled with one or two oil droplets, smooth-walled, 3.2–4.0 × 2.0–2.5 µm (Fig. 7.11).

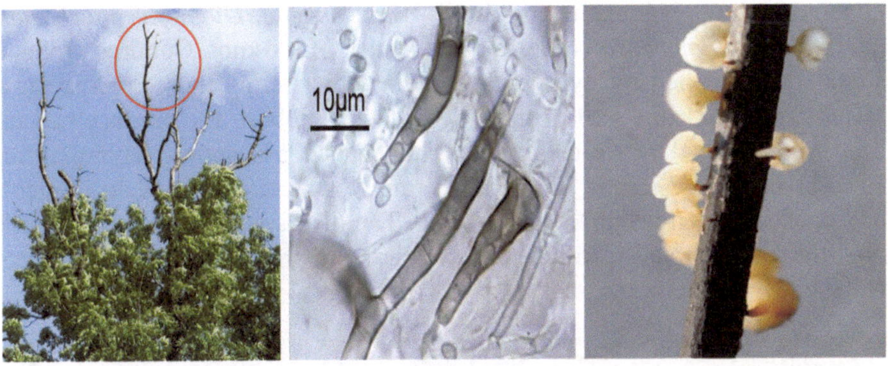

Fig. 7.11: *Hymenoscyphus* die-back on ash, conidia and apothecia.

7.1.12. Monilinia

Monilinia genus belongs to the family *Sclerotiniaceae*. *Monilinia* spp. causing brown rot produce severe losses worldwide in stone fruit production. There are three main fungal species i.e. *Monilinia laxa, M. fructigena*, and *M. fructicola*. A fungal culture is colorless to white, margin entire, aerial mycelium initially sparse, later concentrically zonate. Sporogenous areas buff. Sclerotia present or absent, discoid to elongate, flattened, outer rind dark, 3–4 cells thick, cell walls thick, pigmented, medulla colorless, hyphae intertwined, 2–11 µm diameter, inter-hyphal spaces filled with gelatinous material. The isolates of *M. polystroma* produced similar cylindrical or lemon-shaped conidia and produced two germ tubes per conidium, these germ tubes are generally formed from the pointy sides of the conidia (Fig. 7.12).

Fig. 7.12: *Monilinia* blossom blight on stone fruit and conidia.

7.2. Stem Galls or Swollen Stem

7.2.1. Dibotryon

Dibotryon genus belongs to the family *Venturiaceae. Dibotryon morbosum* or *Apiosporina morbosa* is the causal agent of black-knot. Cultures produce single- or two-celled and cylindrical conidia with tapering ends. Conidia had a mean size of 4.51 μm × 10.97 μm. Pseudothecia globose and produce clavate asci with unequally two-celled and club-shaped ascospores having mean sizes of 6.68–7.37 μm × 16.42–16.78 μm. It affects cherry, plum, apricot, and chokecherry trees in North America. The disease produces rough, black growths that encircle and kill the infested parts and provide a habitat for insects (Fig. 7.13).

Fig. 7.13: *Dibotryon* stem gall on a cherry plant, asci, and ascospore.

7.2.2. Protomyces

The *Protomyces* show a budding, yeast-like growth on laboratory media but do not develop hyphae or sexual states unless infecting their host plants. Species of *Protomyces* generally have a reddish, salmon like color in culture. *Protomyces macrosporus* has a complex life cycle including ascospores and chlamydospores (Fig. 7.14).

Fig. 7.14: *Protomyces* stem gall on coriander and chlamydospores.

8

Identification of Fungi Causing Diseases in Inflorescence and During Grain Formation

8.1. Malformation

8.1.1. Fusarium

The classical example of *Fusarium* infection is *Fusarium moniliforme* which induces malformation in inflorescence, causing mango malformation disease.

The fungal mycelial morphology may vary significantly on different media, and the descriptions here are based upon growth on potato flakes agar (PFA) at 25° C with on and off fluorescent light cycles of approximately 12 h each. The mycelial colony has rapid growth. Colonies are initially white, becoming tinged with lavender, with a colorless to dark purple reverse. Sporodochia, when formed, are cream to orange; however, they are generally sparse on PFA. Dark blue sclerotia may be present.

Microscopically, the fungal hyphae are septate and hyaline. Conidiophores are medium-length (shorter than in *F. solani* and longer than those seen in *F. oxysporum*), simple, or branched. Conidiogenous cells are monophialides. Macroconidia are sparse, very slightly sickle-shaped to nearly straight, i.e. "string bean-like," five-septate, measuring 31–58 × 2.7–3.6 µm. Microconidia are abundant, 0 to 1-septate, oval to clavate, measure 7–10 × 2.5–3.2 µm, and occur in both false heads (a collection of conidia at the tip of the phialide) and chains. Chlamydoconidia are absent.

Fusarium moniliforme differs from the two most common species, *F. solani* and *F. oxysporum*, by forming microconidia in chains and from *F. proliferatum* by lacking polyphialides (phialides with more than one opening not delimited by a septum) (Fig. 8.1).

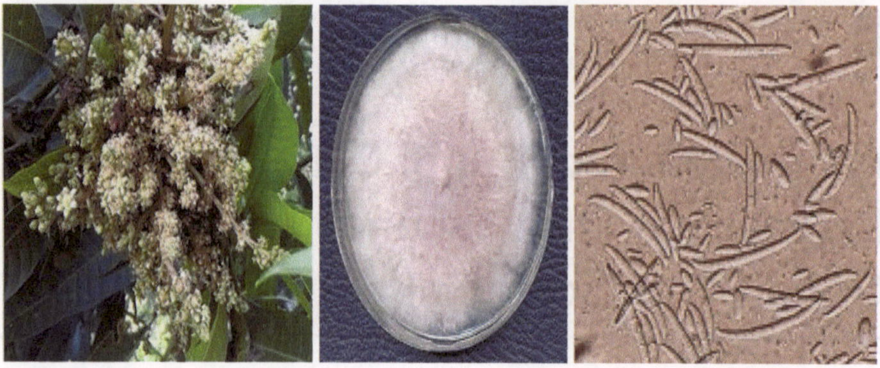

Fig. 8.1: Malformation of inflorescence in mango and *Fusarial* mycelial colony and conidia.

8.2. Sugary disease

8.2.1. Claviceps

Claviceps purpurea (on sorghum) or *C. fusiformis* (on pearlmillet) is an ergot fungus that infects the floret and ovaries in the ear-head of cereal and forage plants to convert these into horn-like ergot sclerotial bodies. The perithecia is flask-shaped with slender and cylindrical asci. The sclerotia is yellow-gray with a rough surface that turns black in due course to replace the infected ovaries or grains. The perithecia of *Claviceps purpurea* is flask-shaped, containing ascus and ascospores (Fig. 8.2).

Fig. 8.2: Sugary exudate of *Claviceps* infection on sorghum and pearl millet with ergot sclerotial bodies, and perithecia with asci.

8.3. Smut

The smuts are multicellular fungi characterized by their large numbers of teliospores. The smuts get their name from a German word for dirt with a dusty black color. This dusty black mass contains dark, thick-walled, and dust-like teliospores. They belong mostly to Ustilaginomycetes (phylum Basidiomycota) and cause smuts in members of the grass family (Poaceae) and sedges (Cyperaceae). Economically important hosts include maize, barley, wheat, oats, sugarcane, and forage grasses. They eventually hijack the plants' reproductive systems, forming galls that darken and burst, releasing fungal teliospores that infect other plants nearby. Before infection can occur, the smuts need to undergo successful mating to form dikaryotic hyphae (two haploid cells fuze to form a dikaryon). The important fungal genera include:

8.3.1. Ustilago

Ustilago is the largest genus in *Ustilaginales*, with approximately 300 species, and most species parasitize monocotyledonous hosts. Most attack the ovaries of Gramineae, and develop in them and the grain to form smut sori, containing the mass of teliospore. The black, dusty teliospores are covered by a membrane of host origin, which ruptures at maturity to release the smut teliospore. These teliospores overwinter (as diploid spores) in soil and on crop debris and can remain viable for several years. The teliospores germinate to form a promycelium, on which a basidium and basidiospores (sporidia) form, following meiosis. *Ustilago* is autoecious, i.e. it completes its life cycle on a single host. Sex organs are completely absent. It produces two kinds of spores during its life cycle, i.e. teliospores and basidiospores. Teliospores of *Ustilago* spp. typically germinate with a cylindrical, mostly four-celled basidium. Each basidial cell repeatedly buds off basidiospores (=sporidia), which produce the haploid yeast state. Teliospore germination readily occurs in many species, or it can be induced by treatments such as soaking in water or plant extracts. Basidium maturation in several species occurs without sporidial development, and in that case the four basidial cells conjugate directly in two pairs. Teliospores are small and formed singly.

The morphology of the teliospore of smut fungi is an important basis for their taxonomic differentiation. The critical taxonomic characters of the smut are the reticulation of the teliospore wall and their sheath configuration. The teliospores of *Ustilago tritici* are 3.6×4.8–5.2 µm and had small pointed regularly spaced surface spines and had small depressions on their sides which appeared spherical, oblong, or cup-shaped (Fig. 8.3).

Fig. 8.3: *Ustilago* smut infection on wheat and teliospore.

8.3.2. Sporisorium

Sporisorium (earlier known as *Sphacelotheca*) causes important diseases like covered kernel smut disease in sorghum plant species all around the world, such as *Sorghum bicolor* (*S. vulgare*), *S. sudanense* (Sudan grass), *S. halepense* (Johnson grass), *Sorghum vulgare* var. *technichum* (broom corn), whip smut in sugarcane, and head smut in maize. It is a seed-borne fungus that grow in the developing shoot. Teliospores appear black to the naked eye and brown under the microscope. They are roughly spherical and 5.5–7.5 μm in diameter. Germination gives rise to a four-celled promycelium that, in turn, produces either hyphae or sporidia from each of the cells. Sporidia continue to multiply asexually by budding in a yeast-like manner. Single-cell colonies continue to grow in a yeast-like manner and may be maintained indefinitely *in vitro*.

Ustilago and *Sporisorium* genera are closely related, but descriptions made by Link, Langdon, and Fullerton allow for a taxonomic boundary of these genera based on variation in columellae, sterile cells, and spore balls and species can be transferred within Ustilago to Sporisorium following this differentiation (Fig. 8.4).

Fig. 8.4: *Sporisorium* infection in sorghum and teliospores.

8.3.3. Farysia

It infects some, but not all, female flowers of *Carex riparia* and *C. pendula* within the inflorescence, producing masses of olivaceous brown spores and long, pale, straw-colored elaters conspicuously projecting from the female spikes, making this one of the more easily identifiable smut fungi in the field. Spores are 6–11 × 4–6 μm and are finely verruculose. Characteristically, the mass of spores is traversed by bundles of hyphae that, after the spores have been blown away, are left behind as a small brush (Fig. 8.5).

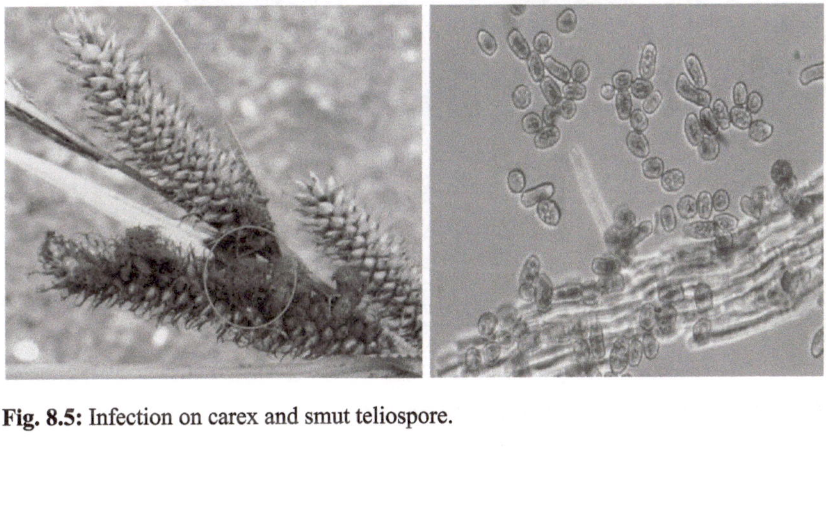

Fig. 8.5: Infection on carex and smut teliospore.

8.4. False Smut

8.4.1. Ustilaginoidea

Ustilaginoidea causes false smut in rice. Individual rice grains are transformed into a mass of yellow fruiting bodies that contain the growth of velvety spores that enclose floral parts. Infected grain has greenish-smut balls with a velvety appearance. The smut ball appears small at first and grows gradually up to the size of 1 cm. Only a few grains in a panicle are usually infected, and the rest are normal. As the fungi growth intensifies, the smut ball containing chlamydospores bursts and becomes orange, then later yellowish-green or greenish-black in color. Infection usually occurs during the reproductive and ripening stages, infecting a few grains in the panicle and leaving the rest healthy.

Chlamydospores formed on the spore balls are born laterally on minute sterigmata on radial hyphae and are spherical to elliptical, warty, olivaceous, 3–5 × 4–6 μm. Younger spores are smaller, paler, and almost smooth. Some of the green spore balls develop one to four sclerotia in the center. These sclerotia overwinter in the field and produce stalked stromata the following summer or autumn. In temperate regions, the fungus survives the winter by means of sclerotia as well as chlamydospores. It is believed that the primary infections are initiated mainly by the ascospores produced from the sclerotia. Chlamydospores play an important role in secondary infection, which is a major part of the disease cycle (Fig. 8.6).

Fig. 8.6: False smut and chlamydospores of the pathogen.

9

Identification of Fungi Causing Fruit Diseases

9.1. Fruit Spot and Blisters

9.1.1. Colletotrichum

The pathogen forms papery or watery brownish areas at the site of infection, with a large number of acervuli on the infected parts. In the acervulus, the conidiophores are slightly curved, short, and club-shaped. The conidia are hyaline and falcate, borne singly on the conidiophores. Numerous black-colored and thick-walled setae are also produced in the acervulus. In culture, most isolates produce cottony, fluffy, or suppressed colonies (Fig. 9.1).

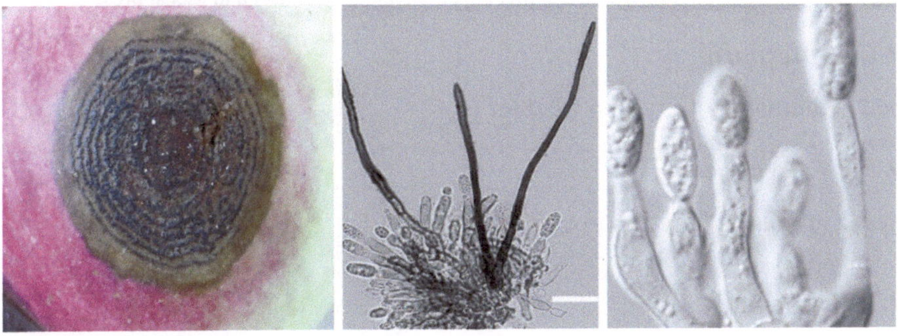

Fig. 9.1: *Colletotrichum* infection on fruit, acervulus and conidia.

9.1.2. Diplodia

See "Diplodia" section in Chapter 6.

9.1.3. Venturia

Venturia is an ascomycete fungus producing sexual spores (ascospores) in a sac-like structure called an ascus (plural asci). The pathogen causes scab disease in the apple by its species *Venturia inaequalis*. The mycelium of *V. inaequalis* is septate, and the nuclei are haploid. A pseudothecium has small dark hairs around its opening and contains pseudo paraphyses along with asci. The asci contain eight haploid ascospores. Asci is an elongated, sac-like structure, each of which contains eight ascospores in a linear arrangement. The ascospores are brown, two-celled spores, and have a characteristic "footprint" shape. The shape of ascospores inspired the Latin name for apple scab, "inaequalis," which refers to the unequal size of the cells. Ascospores measure between 5 and 7 μm wide and between 11 and 15 μm long (Fig. 9.2).

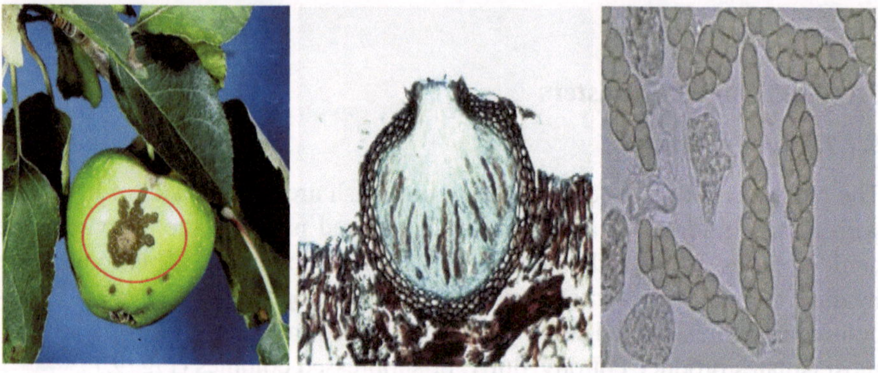

Fig. 9.2: *Venturia* scab infection on apple, Pseudothecium with asci and ascospores.

9.1.4. Alternaria

See "Alternaria" section in Chapter 6.

9.1.5. Curvularia

See "Curvularia" section in Chapter 6.

9.2. Fruit Rot

9.2.1. Botrytis

Botrytis is a genus of anamorphic fungi in the family *Sclerotiniaceae*. It is a plant parasite as well as a saprophyte on both agricultural and forest trees. It produces stout, dark, branching conidiophores that bear clusters of paler conidia (gray in mass) on denticles from the apical ampullae. *Botrytis cinerea* is characterized by abundant hyaline conidia (asexual spores) borne on gray, branching tree-like conidiophores. The fungus also produces highly resistant sclerotia as survival structures in older cultures. It overwinters as sclerotia or intact mycelia, both of which germinate in spring to produce conidiophores. The conidia, dispersed by wind and rainwater, cause new infections (Fig. 9.3).

Fig. 9.3: *Botrytis* gray mold on strawberry and conidia borne on conidiophore.

9.2.2. Aspergillus

Aspergillus causes a disease called "black mold" on certain fruits and vegetables. It forms filamented hyaline and septate hyphae. The conidial heads appear radial and they split into columns (biseriate), the conidiophore vesicle produces sterile cells known as metulae which support the phialides on the conidiophores. The conidiophores are 400–3000 µm long, they are smooth and hyaline later becomes dark at the apex and terminating in a globose vesicle which is 30–75 µm in diameter. The fungal growth has a cottony appearance in the culture plate, initially white to yellow and then turning black comprising conidiophores and conidia that are brown. *Aspergillus niger* is an important species causing rot in fruits and vegetables (Fig. 9.4).

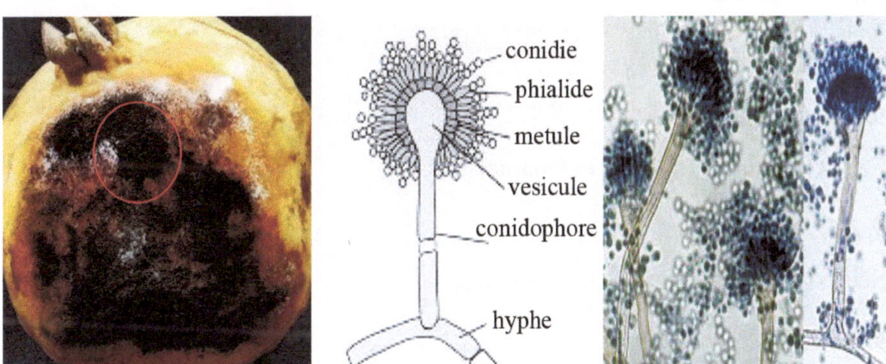

Fig. 9.4: *Aspergillus* fungal colonies and microscopic structure of *Aspergillus*.

9.2.3. Penicillium

Penicillium is an ascomycetous fungi that is of major importance in food spoilage. Fungal mycelial colonies are usually fast growing, in shades of green, sometimes white, mostly consisting of a dense felt of conidiophores. Microscopically, chains of single-celled conidia are produced in basipetal succession from a specialized conidiogenous cell called a phialide. Phialides may be produced singly, in groups, or from branched metulae, giving a brush-like appearance (a penicillus). All cells between the metulae and the stipes of the conidiophores are referred to as branches. Phialides are usually flask-shaped, consisting of a cylindrical basal part and a distinct neck. The important species is *Penicillium italicum* causing rots in fruits (Fig. 9.5).

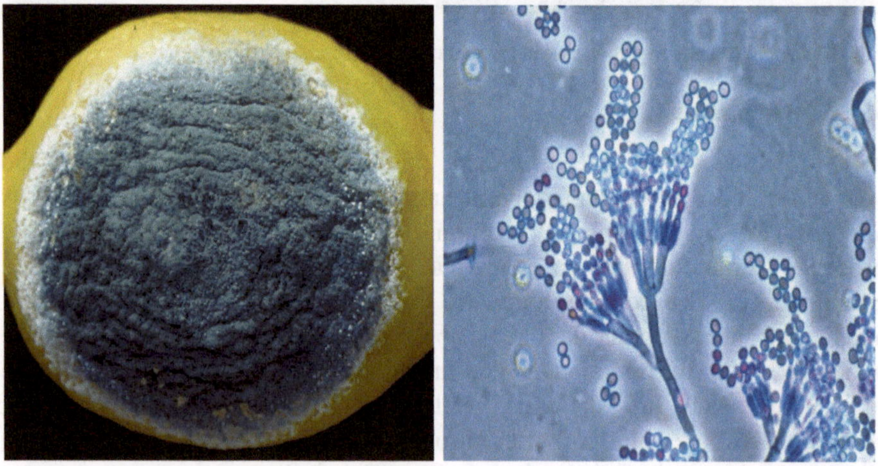

Fig. 9.5: *Penicillium* blue mold rot on citrus.

9.2.4. Curvularia

See "Curvularia" section in Chapter 6.

9.2.5. Alternaria

See "Alternaria" section in Chapter 6.

9.2.6. Colletotrichum

See "Colletotrichum" section in Chapter 9.

9.2.7. *Rhizopus*

Rhizopus is mostly a weak fungal pathogen infecting the post-harvest, ripened, young fruits and vegetables causing soft rot. The fungi are characterized by a body of branching mycelia composed of three types of hyphae, stolons, rhizoids, and usually unbranching sporangiophores. The mycelium is distinguishable into aerial hyphae called stolons, root-like rhizoids, which remain submerged in the substratum. Rhizoids are useful for anchoring the thallus into the substratum and for the absorption of nutrients. Asexual reproduction is by non-motile sporangiospores. Sexual reproduction is by isogametangial copulation. Zygospores are thick-walled, dark, warty sexual resting spores that develop in a zygosporangium. Colonies are fast-growing, covering the surface of the agar. Rapidly growing colonies fade from white to dark during sporulation. The colonies have a dense cottony growth or candy flossy or fairly floss in texture (Fig. 9.6).

Fig. 9.6: *Rhizopus* infection on jackfruit and fungal structure.

9.2.8. *Myrothecium*

Myrothecium species are saprobes in the soil or weak plant pathogens. *Myrothecium* spot is caused by *Myrothecium roridum*. Raised black sporodochia develop on diseased tissue, sporodochia are encircled by a tuft white growth. Initial colonies of *Myrothecium* are white with floccose mycelium, and developed dark green to black concentric rings of sporodochia bearing viscid spore masses. Conidiophores are branched repeatedly. Conidiogenous cells in whorls of three to seven on ultimate branches are hyaline, cylindrical, and 10–14 × 1.5–2 μm. Conidia are hyaline and cylindrical with both rounded ends, occasionally one blunt end, and 6.0–7.6 × 2.2–2.8 μm (Fig. 9.7).

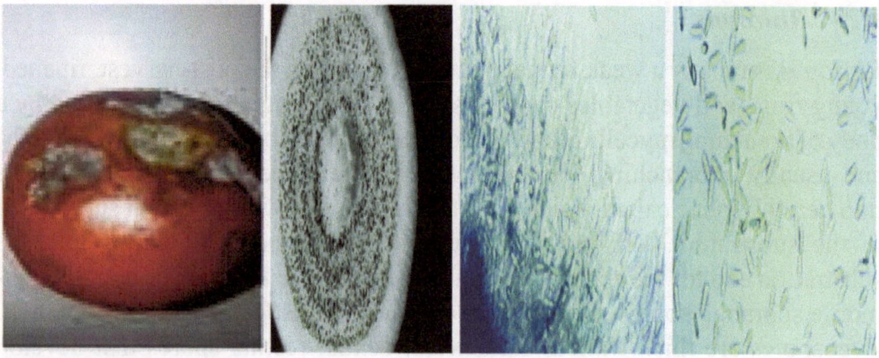

Fig. 9.7: *Myrothecium roridum* on tomato fruit, fungal culture, and conidia.

9.2.9. Pythium

Mycelium slightly branched at the tip of aerial hyphae, hyaline, 8 μm in diameter and non-septate (coenocytic). Sporangia are filamentous, inflated, terminal complex of swollen hyphae developed laterally, zoospores 2–20 per vesicle, reniform in vesicle, 8.0 × 8.1 μm, biciliate laterally, oogonia terminal, spherical, straight, stalked. Antheridia are 1 or 2 intercalary, cordiculate at septa. Thick-walled oospores are solitary, hyaline, thin, aplerotic, spherical, and 10–18 μm in diameter. In species with spherical sporangia, sporangia are generally larger than oospores. Sporangia in *Pythium* produces oospores and zoospores as sexual and asexual spores, respectively. Colonies on PDA are diffuse with cottony aerial mycelium (Fig. 9.8).

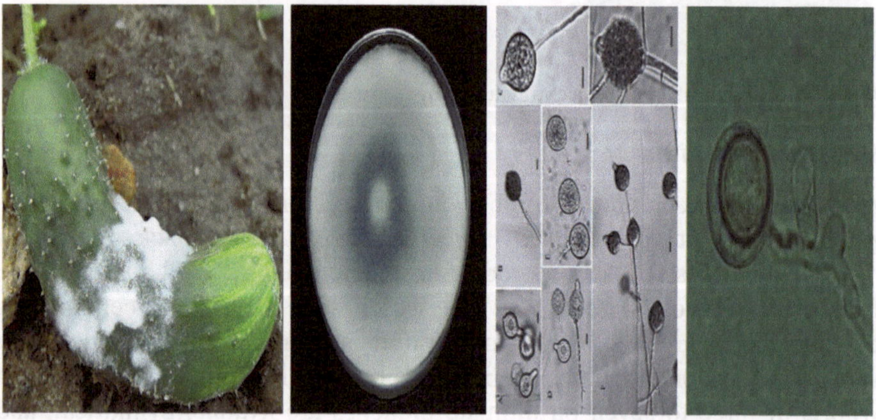

Fig. 9.8: *Pythium* cottony rot of cucumber, fungal culture, sporangium, and oogonium.

9.2.10. Phomopsis

See "Phomopsis" section in Chapter 7.

9.2.11. Rhizoctonia

Rhizoctonia is a genus of anamorphic fungi. Species do not produce spores but are composed of hyphae and sclerotia (hyphal propagules) and are asexual states of fungi in the genus *Thanatephorus*. The hyphae are 4–15μm wide and tend to branch at right angles. A septum near each hyphal branch and a slight constriction at the branch are diagnostic features of this genus. It produces white to deep brown mycelium on an artificial medium (Fig. 9.9).

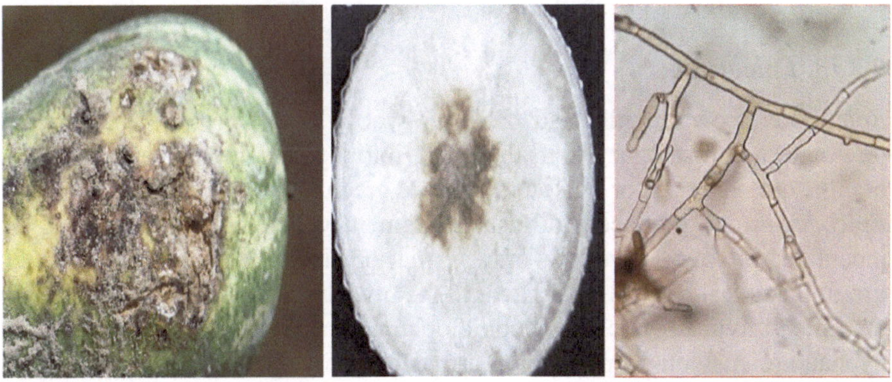

Fig. 9.9: *Rhizoctonia* fruit rot on melon, mycelium with sclerotial bodies and branching of mycelium at right angle.

9.2.12. Diaporthe

Diaporthe is a genus of endophytic filamentous fungal plant pathogens. The mycelium of the pathogen is septate and hyaline, becoming dark with age. Pycnidia are submerged and later become erumpent with a prominent ostiole. Conidia are produced on simple to branched conidiophores and are of two types; alpha conidia, which are subcylindrical, and beta conidia, which are filiform and curved. The perfect stage produces perithecia, in which asci with 8 hyaline, bi-celled, ellipsoid-fusoid ascospores are produced, which are usually constricted at the septum. *Diaporthe* spp. produces a white and cottony mycelial colony with an irregular margin and concentric rings with spherical, black pycnidia scattered across the colony on PDA and exuding cream-colored droplets of conidia (Fig. 9.10).

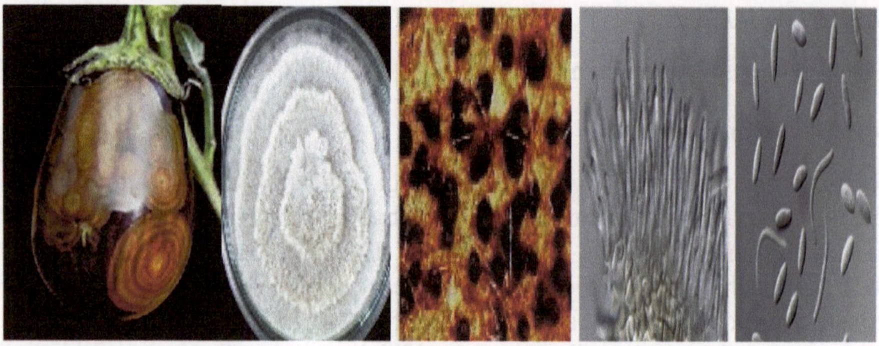

Fig. 9.10: *Diaporthe* infection on brinjal fruit, mycelial growth, pycnidia, and conidia.

9.2.13. Epicoccum

Epicoccum grows rapidly and produces wooly to cottony or felty colonies on potato dextrose agar. From the front, initially the colonies are yellow to orange, orange to red or pink, and become greenish brown to black by aging. *Epicoccum* may produce a diffusable pigment which turns the color of the inoculated medium to yellow, orange, red, or brown. Black dots (100–2000 μm in diameter) may be visualized on the mycelial colony surface. These are the tufts of hyphae which have conidiophores on their surface. These tufts of hyphae are cushion-shaped and non-convoluted and are called sporodochia. The hyphae are septate and yellow to brown in color. The conidia are usually produced in dense masses from clustered conidiophores (sporodochia). Colonies are fast growing, suede-like to downy, with a strong yellow to orange–brown diffusible pigment. When sporulating, numerous black sporodochia (aggregates of conidiophores) are visible (Fig. 9.11).

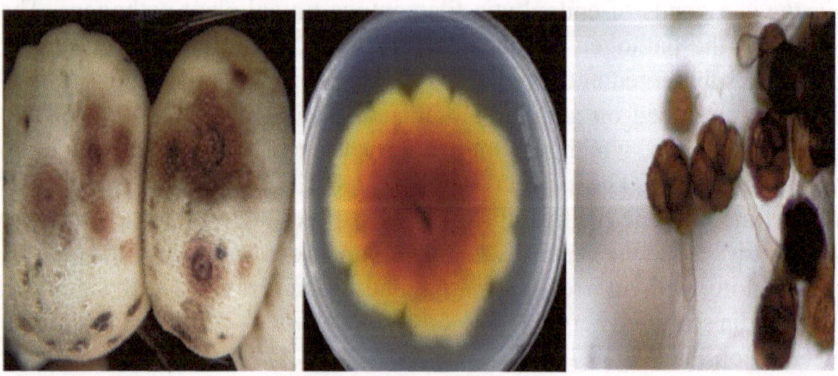

Fig. 9.11: *Epicoccum* infection on potato, mycelial growth, and spores of the fungus.

9.2.14. *Trachysphaera*

Trachysphaera fructigena cause cigar end rot in banana. Conidiophores are usually solitary or in small groups. Conidia are hyaline, oblong to cylindrical. They are borne at the ends of tapering phialides, aggregated into round heads (Fig. 9.12).

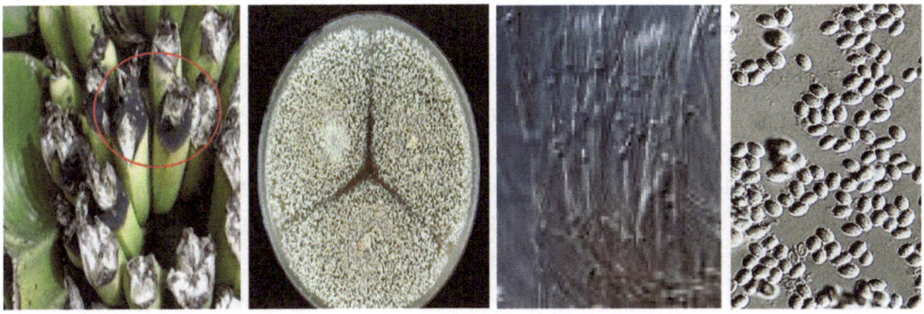

Fig. 9.12: *Trachysphaera* infection on banana, the fungus growth, conidiophore, and conidia.

9.2.15. *Verticillium*

Verticillium genus belong to the division *ascomycota*. The vegetative mycelium is hyaline, septate, and multinucleate. Conidiophores are usually well differentiated and erect, verticillately branched over most of their length, bearing whorls of slender awl-shaped divergent phialides. Conidia are hyaline or brightly colored, mostly one celled, and are usually borne in slimy heads. Thick-walled resting structures are formed known as microsclerotia, which are elongated or globose (Fig. 9.13).

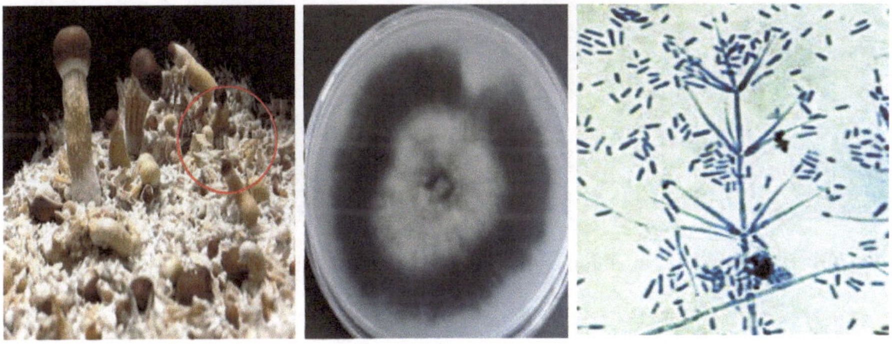

Fig. 9.13: *Verticillium* infection on mushroom, mycelial growth, and conidiophore structure with conidia.

9.2.16. Marasmiellus

The fungus causes coconut embryo rot. Infection by *Marasmiellus inoderma* occurs while the seed-nuts are attached to the coconut palm. Most likely, it occurs as airborne spores, which enter through the stalk (or calyx) end of the seed-nut, colonize the fibrous husks and grow beneath the brown cap as it is raised for germination. The embryo is destroyed, and the fungus enters the nut cavity forming a pinkish-white fungal growth floating on the liquid endosperm milk. Foul-smelling secondary rots often develop.

Marasmiellus is a *Basidiomycete* fungus in the order *Agaricales*. Pure culture of *Marasmiellus* produces mycelial colonies that are fast growing on PDA and MEA media with dense light mycelia with lighter orange color at the center. The binucleate hyphae are mostly not immersed in agar. Clamp connections are abundantly present in all isolates. The morphology of basidiocarps and basidiospores is similar with differences in the size and color of basidiocarps and size of basidiospores. The fungus pileus is pale orange in the middle with a depressed center and measures 9–30 mm in diameter. The stipe is white, pale orange, and silky, the base is initially filled with tissues then becomes hollow, bulbous, and eccentric and measures 7–29 mm in diameter. The basidiospores are hyaline, oblong to ellipsoid, and smooth (Fig. 9.14).

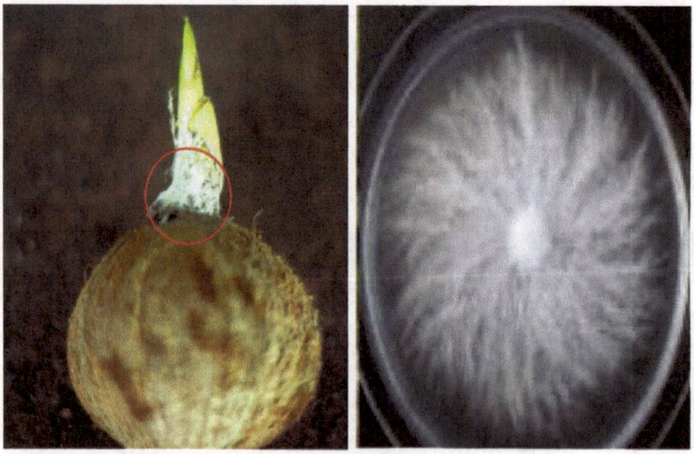

Fig. 9.14: *Marasmiellus* infection on coconut seed-nut and fungal pathogen.

9.2.17. Nigrospora

Nigrospora causes the center of the banana fruit to turn dark red. *Nigrospora* can infect the fruit in tropical climates where bananas are grown. *N. sphaerica* colonies grow rapidly and appear hairy or wooly. The conidiophores are short and clustered, surfacing from mycelium. They appear translucent in color and have an average range of 8–11 μm in diameter. The conidiophores are often straight or slightly curved. Conidia grow from the tips of the translucent conidiophores. The conidia are brownish black, oblate spheroid, and single-celled. On average they range from 16 to 18 μm in diameter. The initial white, translucent-looking colony of *N. sphaerica* turns brown or black due to the mass sporulation of conidia from the conidiophores. In laboratories, *N. sphaerica* is grown on potato dextrose agar (PDA) at room temperature. The spore is transparent at this time and is observed to have a tiny projection at the tip, which is observed after the spore became quite black. The mature spore shows no evidence of this projection and appears to be quite smooth. Conidiophores arose terminally and from branched laterals (Fig. 9.15).

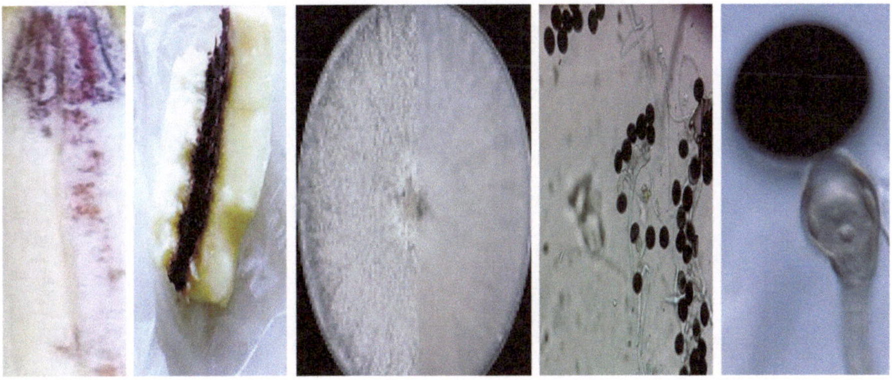

Fig. 9.15: *Nigrospora* on banana, fungal growth, and conidiophore with conidia.

10

Identification of Fungi Causing Grain and Seed Diseases

10.1. Black Tip

The black tip of grain infection is mostly observed on the wheat grains.

The fungal pathogen involved in causing the black tip of grain includes the species of the fungal genus *Alternaria* and *Helminthosporium.*

10.1.1. Alternaria

The genus stands distinct from other genera by its transversely and longitudinally septate (muriform) conidia. The conidia are usually formed in chains. These are dark brown, typically obclavate with a distinct oval body and beak. The beak is light in color, which may be short or long and serves as a conidiophore. *Alternaria* can be isolated well on PDA medium with streptomycin in 3–5 days. The fungal mycelium in isolation plates is usually grey to black (Fig. 10.1).

Fig. 10.1: Black tip infection on wheat grain and conidia of fungus *Alternaria.*

10.1.2. Helminthosporium

The hyphae are typically dark, bearing conidia on simple, erect, septate conidiophores. The conidia are obclavate, brown, and thick-walled with 8–10 transverse septa and a basal scar, indicating the point of attachment on the conidiophore. The conidia are produced singly from a pore at the tip and on lateral sides in the conidiophore's upper part, often forming verticles (whorls) above or below the septa.

Helminthosporium culture on potato dextrose agar (PDA) is olivaceous brown to black. In old cultures, the number of conidia is so abundant that the entire culture turns black and shiny (Fig. 10.2).

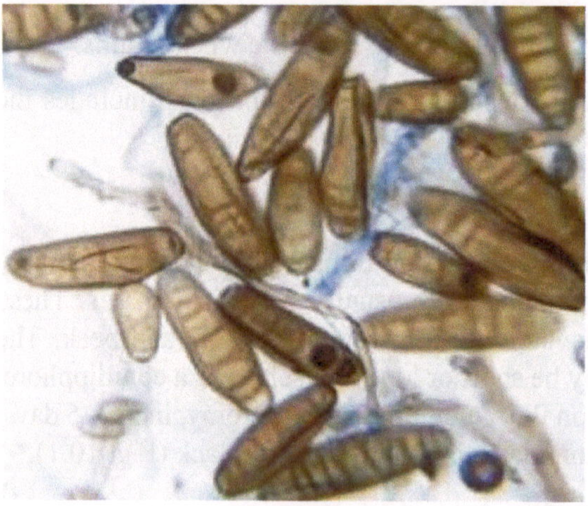

Fig. 10.2: Conidia of black tip fungus *Helminthosporium.*

10.2. Bunt

Bunt, also known as stinking smut, is a disease of grains in wheat, rye, and other grasses caused by *Tilletia*. Infection by *Tilletia tritici* (formerly *T. caries*) or *T. laevis* (formerly *T. foetida*) causes normal kernels to be replaced by "smut balls" containing powdery masses of brownish black spores characterized by a dead-fish odor.

Tilletia caries is a basidiomycete that causes common bunt of wheat. The common names of this disease are stinking bunt of wheat and stinking smut of wheat. This pathogen infects wheat, rye, and various other grasses. *T. caries* is economically and agriculturally important because it reduces both the wheat yield and grain quality.

10.2.1. Tilletia

Tilletia is a genus of smut fungi in the *Tilletiaceae* family. The species in this genus are plant pathogens that affect various grasses. Teliospores are one-celled, dusty, and escape at maturity. Spores are large and are reticulate and verrucose (warts). Sporidia are fused to form H-shaped structures.

The traditional identification of *Tilletia* species is based on the morphological features of the spores. Microscopic observations show that *T. laevis* has a smooth teliospore cell wall surface, whereas the surface of *T. caries* is reticular. The mean diameter of *T. caries* teliospores is 14–23.5 µm, occasionally 25 µm. Similarly, the diameter of *T. laevis* teliospores ranges from 13 to 22 µm. *T. controversa*, causing dwarf bunt in wheat, has similar morphology and genetic structure to *T. caries*. Teliospores of *T. tritici* have ridges on their walls, which gives them a "warty" appearance, while teliospores of *T. laevis* have smooth spore walls. The teliospores of *T. tritici* are thick-walled, globiose, reticulate, and 13–23 µm in diameter.

Tilletia indica teliospores are globose to sub-globose, sometimes with a small hyphal fragment (more common on immature teliospores, but occasionally found on mature teliospores); mostly 22–47 µm in diameter, occasionally larger, up to 64 µm (mean 35–41 µm); pale orange-brown to dark reddish brown; mature teliospores are black and opaque, densely ornamented with sharply pointed to truncate spines, occasionally with curved tips, 1.4–5.0 µm high, which in surface view appear as either individual spines (densely echinulate) or closely spaced, narrow ridges (finely cerebriform); the spines are covered by a thin hyaline membrane. Sterile cells of *T. indica* are globose, sub-globose to lachrymiform (tear-shaped), yellowish brown, 10–28 µm × 48 µm, with or without an apiculus (short stalk), with smooth walls up to 7 µm thick and laminated.

Tilletia walker mean spore size is 30–31 µm, max spore size is 36–45 µm with pale yellow to dark reddish -brown colored spore with globose shape. The spore ornamentation is coarse, cerebriform, and irregular with gaps between spines obvious in profile after bleaching.

Tilletia horrida mean spore size is 24–28 µm, max spore size is 36 µm with pale yellow-dark or light chestnut brown spore with globose to sub-globose shape. The spore ornamentation is echinulate, sometimes cerebriform ridges. Generally curved and polygonal scales.

Tilletia ehrhartae mean spore size is 24–28 µm, max spore size is 28 µm with very dark olivaceous brown (mature spore), opaque (melanization of the scale) with globose to sub-globose shape. The spore ornamentation is cylindrical to slightly tapered spines and rarely cerebriform; broadly truncated to slightly rounded at the apex (Fig. 10.3).

Fig. 10.3: Bunt and its teliospore.

10.3. Smut or Partial Grain Smut

The inflorescence or grains are transformed in to the smut sori. Different genera of the fungus are involved in the smut diseases.

Ustilago and *Sporisorium* genera are closely related, but descriptions made by Link, Langdon, and Fullerton allow for a taxonomic boundary of these genera based on variation in columellae, sterile cells, and spore balls.

Identification of grain smut genera:

A key to the genera of the *Ustilago–Sporisorium–Macalpinomyces* complex is given below. The genera can be identified based on soral characteristics, host plant, presence and absence of columella, sterile cell, and spore balls (Mc Taggart et al., 2012. Persoonia 29: 116–132).

1. On *Polygonaceae*............... *Melanopsichium*
2. On *Triodia*............... *Triodiomyces*
3. On other genera of grasses............... *Ustilago*
4. Columella filiform, flattened, or flexuous. Spore balls formed from coiled sporogenous hyphae. Sterile cells are usually absent............... *Anthracocystis*
5. Columella stout, cylindrical or woody. Spore balls are usually absent. Sterile cells present............... *Sporisorium*
6. Sori cylindrical or tubular, derived from host tissue............... *Macalpinomyces*

7. On *Aristida*................ *Langdonia*
8. On *Panicum trachyrachis*, sterile cells present................ *Anomalomyces*
9. On *Eriachne*................ *Macalpinomyces* s.str.
10. On grasses in the tribe *Andropogoneae*................ *Stollia*
11. On grasses in the tribes *Panicoideae* or *Chloridoideae*................ *Macalpinomyces*

10.3.1. Ustilago

Ustilago, the largest genus of the family *Ustilaginaceae* is represented by more than 400 cosmopolitan species. Butler and Bisby (1958) reported 108 species from India. All species are parasitic and infect the floral parts of wheat, barley, oat, maize, sugarcane, bajra, rye, and wild grasses. The mycelium is hyaline but changes to brown near maturity. The mycelial cells are transformed into brown, spherical teliospores which are spherical to ellipsoid, and olive-brown to black in color (Fig. 10.4).

Fig. 10.4: Teliospores of *Ustilago* sp.

10.3.2. Sporisorium (Sphacelotheca)

Grain kernels are replaced by fungal spore-producing structures known as sori. Smut sori are covered by a peridium, a tough membrane that appears grayish-brown in color. The shape of sori, themselves is generally oval or conical. Often, sori appears as an elongated sorghum seed. Variations in the size of sori range anywhere from 0.4 to 1.3 cm long and 0.2–0.4 cm wide. Glumes, or tiny leaves, will sometimes cover very small sori. They appear white, gray, or brown. In some instances, sori may have a striped appearance. Dark brown, powder-like masses of sori may be concentrated on a particular region of an infected head, or in some instances, all kernels of a smutted head can be destroyed. Missing or distorted spikelets covered in sori on the sorghum panicles also indicate signs of infection in rare instances. The *Peridium* of interwoven hyphae is overlain by several layers of host tissue. *Columella* is composed of host tissues permeated by inter- and intra-cellular hyphae, cylindrical, stout or woody, branched or unbranched, and peripheral cells thick-walled and vacuolated. *Hyphae* growing from the columella of young sori differentiate as pockets of sporogenous hyphae enclosed by non-sporogenous partitioning hyphae. Sporogenous hyphae are uncoiled. Spores are at first somewhat agglutinated, later pulverulent, dark, single, and globose to subglobose. Sterile cells are hyaline, sub-globose to globose, in groups or chains, intermixed with the spores, and formed from non-sporogenous partitioning hyphae. The presence or absence of sterile cells depends on the species of *Sporisorium*. Germination is of the *Ustilago* type. The pathogen of the disease in its development cycle forms teliospores (wintering stage of the fungus) and sporidia. Teliospores are spherical or roundish, tinted, smooth, and 4.8–8.5 µm in diameter. At sprouting, they form three or four cellular basidia with laterally and apically growing sporidia. Sporidia are oblong, with slightly rounded ends, or spindle-shaped, 2.9–18.5 µm (Fig. 10.5).

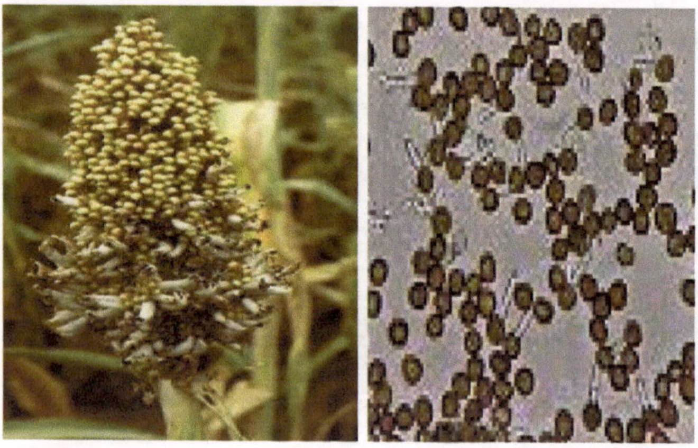

Fig. 10.5: *Sporisorium* infection on sorghum and the fungal Teliospore

10.3.3. Ustilaginoidea

Ustilaginoidea virens causes false smut in rice. Individual rice grains transform into a mass of yellow-fruiting bodies. Infected grain has greenish-smut balls with a velvety appearance. The smut ball appears small at first and grows gradually up to the size of 1 cm. It is seen in between the hulls and encloses the floral parts. Only a few grains in a panicle are usually infected, and the rest are normal. As the fungi growth intensifies, the smut ball bursts and becomes orange, then yellowish-green or greenish-black in color. Infection usually occurs during the reproductive and ripening stages, infecting a few grains in the panicle, and leaving the rest healthy. Chlamydospores formed on the spore balls are born laterally on minute sterigmata on radial hyphae and are spherical to elliptical, warty, olivaceous, 3–5 × 4–6 µm. Younger spores are smaller, paler, and almost smooth. Some of the green spore balls develop one to four sclerotia in the center. These sclerotia overwinter in the field and produce stalked stromata the following summer or autumn. In temperate regions, the fungus survives the winter using sclerotia as well as chlamydospores. It is believed that the primary infections are initiated mainly by the ascospores produced from the sclerotia. Chlamydospores play an important role in secondary infection, which is a major part of the disease cycle (Fig. 10.6).

Fig. 10.6: False smut on rice and teliospores and chlamydospores of *Ustilaginoidea* sp.

10.3.4. Ephelis (Balansia)

Ephelis (Balansia) oryzae causes Udbatta disease in rice. The fungus forms a stroma over the entire length and girth of the inflorescence, and black, convex pycnidia are formed in the stroma. Conidia are colorless, needle-shaped, non-septate, and contain one nucleus per conidium. Colonies on PDA are first white cottony, then turn yellowish or brownish with laciniate margins. Mycelia grow at temperatures of 15–30° C (optimum 20–28° C) and never at 35° C (Fig. 10.7).

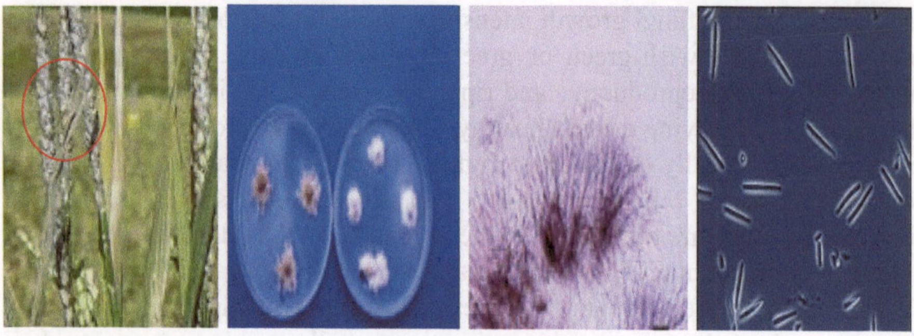

Fig. 10.7: Infection of *Ephelis* on rice panicles and fungal conidia.

10.3.5. Tolyposporium

Tolyposporium causes smut in pearl millet. The sporeballs of *Tolyposporium penicillariae* varied from circular to near-polyhedral and measured 42–325 × 50–175 µm. The number of teliospores aggregated in balls varied from 200 to 1400. Teliospores are mostly circular and measure 7–12 µm diameter. Maximum germination of teliospores aggregated in balls occurred at 30° C and different patterns of germination are observed. Large variability is observed in the cultural characteristics of the pathogen grown on different media. The fungus grows well within 3–5 days on potato or carrot agar at 35° C. Growth of the fungus remains purely sporidial even after repeated subculturing. Sporidia are borne on promycelia, laterally, and terminally, and are reproduced by budding in chains. Sporidia are spindle-shaped and vary in length from 8 to 25 µm. Individually separated teliospores seldom germinate (Fig. 10.8).

Fig. 10.8: *Tolyposporium* infection on pearl-millet and teliospores.

10.4. Ear-head or Cob rot

10.4.1. Diplodia

Diplodia (*Stenocarpella*) ear rot is characterized by a white mold on the ears. Conidia are light brown to dark mature. Pycnidia contained ellipsoid to ovoid, hyaline and aseptate, or dark brown and septate conidia. Pathogens are able to fully colonize an 85 mm diameter petri plate in 48 h (Fig. 10.9).

Fig. 10.9: *Diplodia* infection on maize grain and fungal conidia.

10.4.2. Cephalosporium

Cephalosporium maydis cause infection on maize kernel. Their conidia are usually one-celled, hyaline, or pigmented, globose to cylindrical, and mostly aggregated in slimy heads at the apex of each phialide (Fig. 10.10).

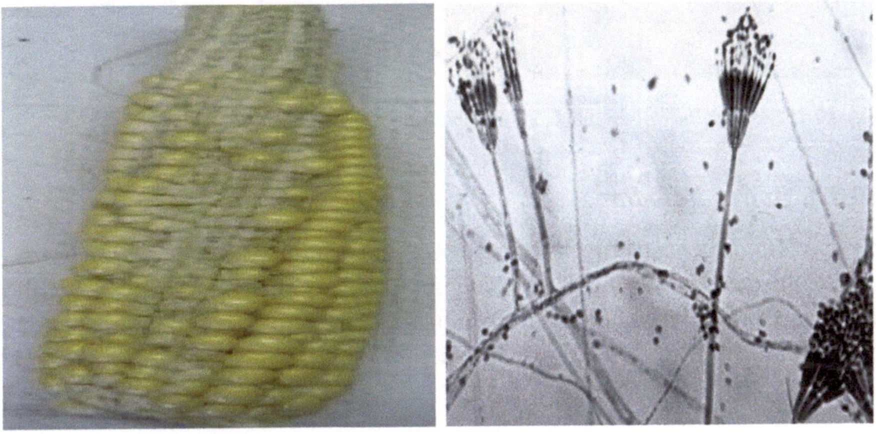

Fig. 10.10: *Cephalosporium* on maize kernel and fungal structure.

10.4.3. Aspergillus

The fungus *Aspergillus flavus* causes *Aspergillus* ear rot, one of the most important diseases in corn. The *Aspergillus flavus* group of fungi has a complex morphology that is classified based on the formation of sclerotia. Group I has L-strains whose sclerotia are greater than 400 µm in diameter, and Group II has S-strains with sclerotia less than 400 µm in diameter. They also have both sexual and asexual forms of reproduction. Asexual reproduction produces conidia and sclerotia, while sexual reproduction produces sclerotia. The conidial spores are produced on the phialides, originating from the conidiophore vesicles. The conidial spores have a size of 3–6 µm. The conidiophore originates from the hyphal threads. The conidiophores are colorless and rough-textured. The phialides originating from the conidiophore vesicles are both uniseriate and biseriate. Hyphae occur as thread-like septate branches that form mycelium. *Aspergillus flavus* can also reproduce sexually by producing ascospores contained within the sclerotia (Fig. 10.11).

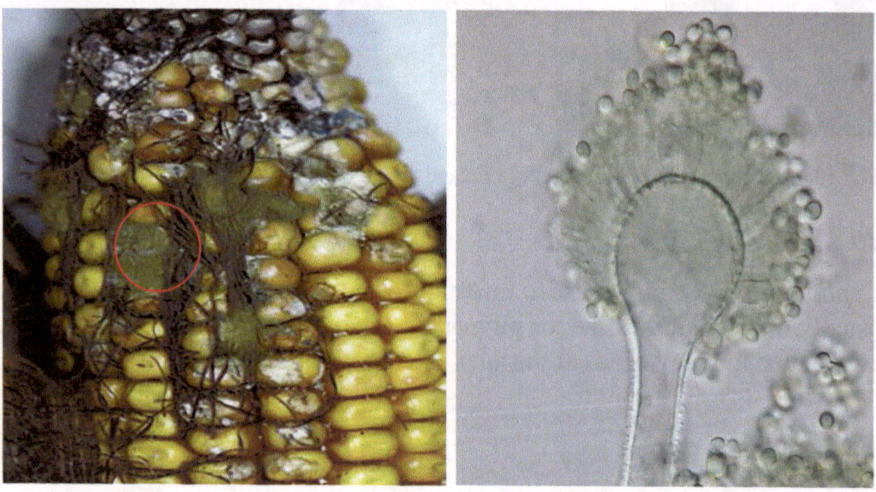

Fig. 10.11: *Aspergillus* rot on maize kernel and fungal structure.

10.4.4. Botryodiplodia

Botryodiplodia causes black kernel rot in maize. *Botryodiplodia theobromae* on potato dextrose agar (PDA) medium grows rapidly at 20–30° C, with a white to light or dark gray aerial mycelia on PDA, and later produces abundant black pigmentation, which can be seen from the reserve side of the PDA culture. *Botryodiplodia theobromae* produces pycnidia, which are sub-globose to globose (300–700 μm in diameter) and single or grouped in a stroma and papillate. Conidiophores are hyaline, simple, sometimes septate, rarely branched cylindrical, and arise from the inner layers of cells lining the pycnidial cavity. Conidia (10–8 × 17–43 μm) are formed in the pycnidia. Young conidial cells are hyaline and granular and have no septum, while mature conidia are striated and contain one septum. Once the two-celled spores are formed, they readily survive from one season to the next. Perithecia of the fungus can be formed singly or in groups. They are globose (225–300 μm in diameter) and papillate and could produce ascospores (11–14 × 30–35 μm) that are hyaline, one-celled, and elliptical to ellipsoidal (Fig. 10.12).

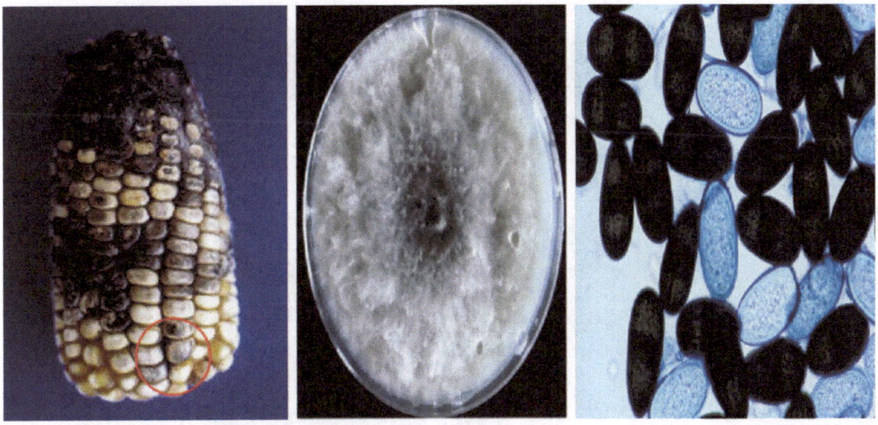

Fig. 10.12: *Botryodiplodia* on maize, fungal growth and conidia.

10.4.5. Cunninghamella

It is a seed-borne pathogen. *Cunninghamella* is a member of the family *Cunninghamellaceae* (phylum *Mucoromycota*). The species *Cunninghamella echinulata* is closely related to *C. elegans*, and both species share highly similar characteristics of growth and morphology. Colonies tend to be rapidly growing on most growth media, producing dense, white, or grayish aerial mycelium. *Cunninghamella echinulata* reproduces asexually and solely *via* yellow-brown, spiny, single-spored sporangioles that, due to the nature of the sporangiospore being retained within the sporangium, appear to have a two-layered outer wall. This fungus grows by means of filaments that lack septa.

This is a common feature of members of the *Mucoromycota,* where the hyphal compartments are either fully divided by septa or are completely continuous (coenocytic) and multinucleate. Zygospores of this fungus are only produced following the fusion of gametangia of compatible mating strains, exemplifying a heterothallic mating system. The sporangiophores of this species are irregularly branched and do not resemble the sporangiospores typical of most other members of the *Mucoromycota* encountered in similar habitats. The sporangioles produced by this fungus are larger (10–20 μm) than those of the closely related species, *C. elegans.* sporangioles and sporangiospores are visualized in the fungal cultures. Sporangiophores are erect and form short lateral branches, each of which terminates in a swollen vesicle terminating in sporangioles (Fig. 10.13).

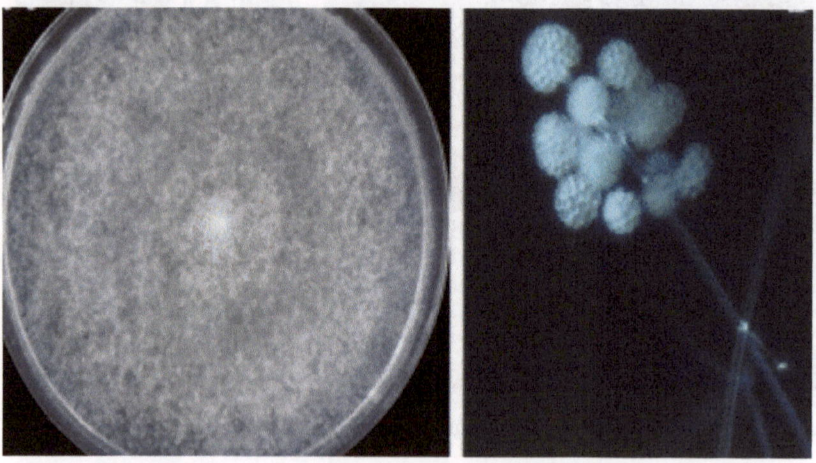

Fig. 10.13: Fungal culture and Sporangioles of *Cunninghamella.*

10.4.6. Fusarium

The fungus causes *Fusarium* ear rot in maize. The hyphae of the fungus are hyaline, branched, and septate. Conidia are formed in slimy, effuse sporodochia called pionnotes, or sometimes scattered on the mycelium. Conidiophores are quite short, simple, branched, or septate, bearing a terminal phialide. The conidia, which are slimy, are of two types, i.e. microconidia (one-celled, oval or comma-shaped, pyriform or elongate) and macroconidia (which are hyaline or pale spindle-shaped with pointed ends). They are two- or more-celled, thick-walled, smooth, and cylindrical or sickle-shaped. Monophialides and polyphialides (in heads or in chains) may be observed (Fig. 10.14).

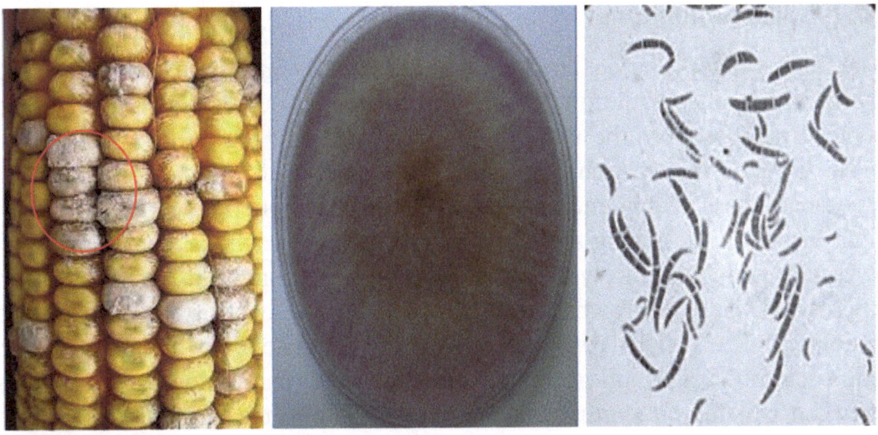

Fig. 10.14: *Fusarium* on maize kernel, mycelium, and conidia.

10.4.7. Gibberella

The pathogen causes *Gibberella* ear rot of maize. In maize, this disease is characterized by symptoms on both ear and stalk. Early symptoms are often visible on the tip of the ear as white mold, which over time turn pink or red. *Gibberella zeae* also known by the name of its anamorph *Fusarium graminearum*, also causes *Fusarium* head blight (FHB) in wheat and barley. The sexual state, *G. zeae*, is characterized by purple to black perithecia—a distinguishing feature of the genus *Gibberella*. The ascospores like the conidia are septate, usually with three septa (Fig. 10.15).

Fig. 10.15: *Gibberella* on maize kernel and fungal ascospores.

10.4.8. Macrophomina

The fungus causes charcoal rot on maize kernel. *M. phaseolina* is characterized by hyaline hyphae with thin walls and light brown or dark brown hyphae with septa. Branches from the main hyphae are generally formed at a right angle to the parent hyphae, with constriction at the point of origin. Microsclerotia, a compact mass of hardened fungal mycelium, are spherical, oval, or oblong, light brown in the young stage, becoming darker (brown to black) with age. Pycnidia, which are rarely observed under natural conditions, are larger than microsclerotia, dark brown to black, rough, globose, or irregular, beaked, and ostiolated. The fungus can show a wide heterogeneity in mycelium color, microsclerotia distribution, pycnidia formation, and chlorate phenotypes between isolates on synthetic media. Nevertheless, the amplification of the internal transcribed spacers (ITS) has indicated that isolates belonged to one single species. It has been suggested that the morphological heterogeneity could be attributed to the responses of the fungi to environmental factors or variations in their host species (Fig. 10.16).

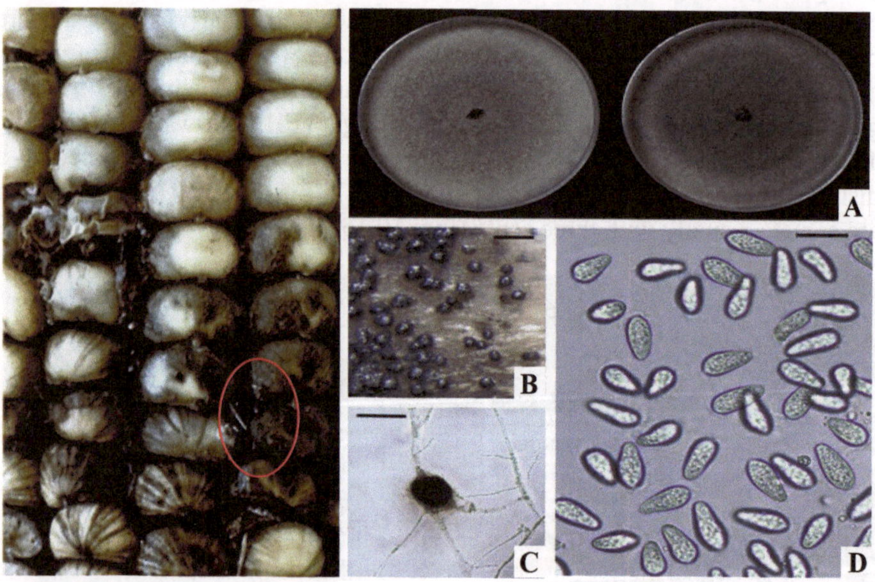

Fig. 10.16: *Macrophomina* charcoal rot on maize kernel; **(A)** fungal colony, **(B)** microsclerotia, **(C)** germinating sclerotia, and **(D)** ascospores.

10.4.9. Rhizoctonia

The fungus *Rhizoctonia* causes infection in potato and seed tubers. *Rhizoctonia* can cause multiple impairments, such as black scurf on tubers, blemishes such as "elephant hide," deformed tubers and cracks, and small lesions not larger than 5 mm. This is the phase of the formation of sclerotia on tubers. Sclerotia are the survival structures of the pathogen. These sclerotia are often referred to as black scurf. The pathogen overwinters as sclerotia on infected tubers, which can be a source of disease infection and spread in new area. *Rhizoctonia* have the following characteristics: (1) some shade of brown; (2) a special type of cross wall within the hyphae, called a dolipore septum; (3) each cell is multinucleate (has many nuclei) rather than binucleate; (4) branches that are produced at right angles; (5) no asexual spores are formed by the mycelium (Fig. 10.17).

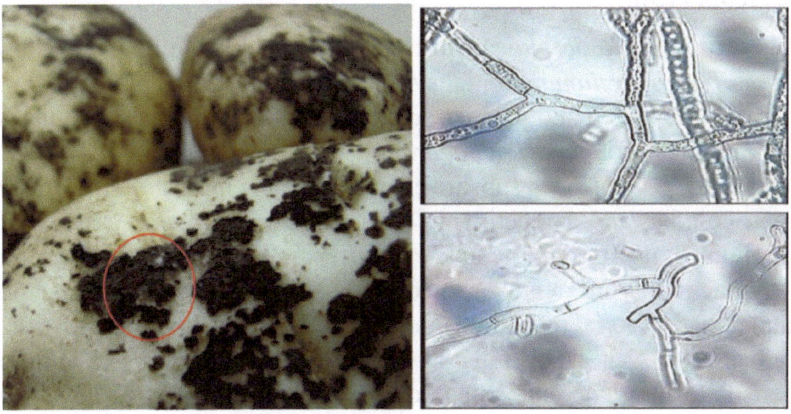

Fig. 10.17: *Rhizoctonia* on potato seed tuber, sclerotia and mycelium structure.

10.4.10. Penicillium

Penicillium ear rot in maize kernels is caused by multiple *Penicillium* spp. It is characterized by green or blue-green powdery mold growing between kernels, usually at the tip of the ear. Infected kernels can appear bleached or streaked. *Penicillium* spp. are initially white and become blue-green, grey-green, olive-grey, yellow, or pinkish with time. *Penicillium* species have septate hyphae (2–5 μm in diameter) that give rise to branched or unbranched conidiophores with secondary branches that give *Penicillium* a brush-like appearance. *Conidia* are unicellular, globose, or ovoid and measure 2.5–5 μm in diameter. They are visualized as chains at the tips of the phialides borne on conidiophores. Sexual reproduction involves the production of asci and ascospores. The asci contain eight unicellular ascospores (Fig. 10.18).

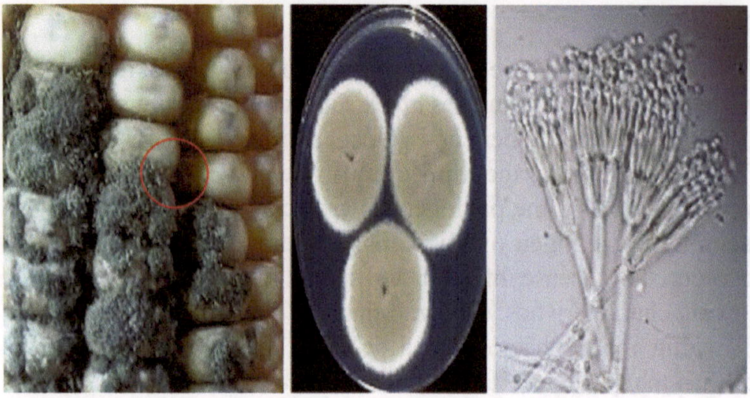

Fig. 10.18: *Penicillium* ear rot of maize, mycelial colonies and fungal structure.

10.4.11. Trichoderma

The fungus causes *Trichoderma* ear rot in maize kernels. The fungal species responsible is *T. afroharzianum.* Initial symptoms include white mold growth, which later on turns green and powdery on and in between kernels and husks often involving the whole ear. *Trichoderma* ear rot damage is usually spotty throughout a field. *Trichoderma* spp. are characterized by rapid growth, mostly bright green conidia, and a repetitively branched conidiophore structure of green mycelium typical of the genus *Trichoderma.* The colonies grown on PDA and synthetic low-nutrient agar medium (SNA) shows green, abundant, cottony, and radiating aerial mycelium and yellow pigmentation on the reverse. Colony radius after 72 h at 30° C is 60–65 mm on PDA and 50–55 mm on SNA. The isolates produce one-celled conidia 2.8–3.8 μm long and 2.1–3.6 μm wide on SNA. Conidiophores and phialides are lageniform to ampulliform and measure 4.5–9.7 μm long and 1.6–3.6 μm wide; the base measure 1.5–2.9 μm wide, and the supporting cell measured 1.4–2.8 μm wide (Fig. 10.19).

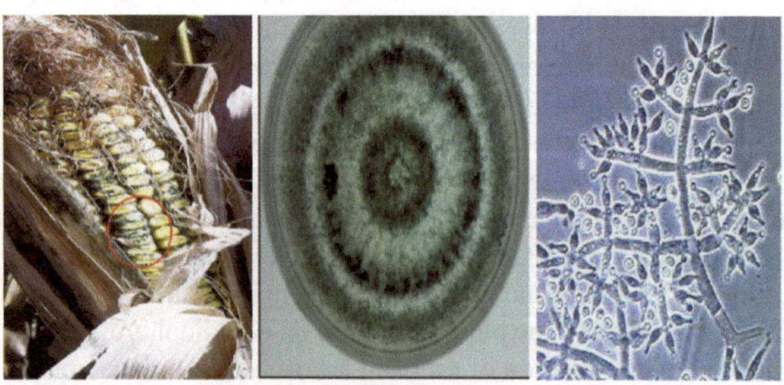

Fig. 10.19: *Trichoderma* ear rot in maize, fungal colony, and structure.

10.5. Grain Mold

Different fungal species causes grain mold particularly in grain of standing crops of cereals and on grains of pulses, oilseeds in post-harvest storage and on dry-fruits (Fig. 10.20).

Fig. 10.20: Grain mold on jowar, maize cob, on stored oilseed, pulse grains and dryfruit.

10.5.1. Aspergillus

The hyphae of the fungus are septate, hyaline, and branched. The conidiophores originate from the basal foot cell located on the supporting hyphae and terminate in a vesicle at the apex. The mature vesicles have phialides over their entire surface, which bear a chain of conspicuously echinulate conidia. Conidia are produced in enormous quantity. The conidia are arranged in chains, with the youngest conidia at the base and the oldest at the top. Vesicle-bearing phialides and chains of conidia are typical characteristics of the genus *Aspergillus*. The morphology and color of the conidiophore vary from one species to another (Fig. 10.21). They are the "green molds" and are also called as "storage fungi."

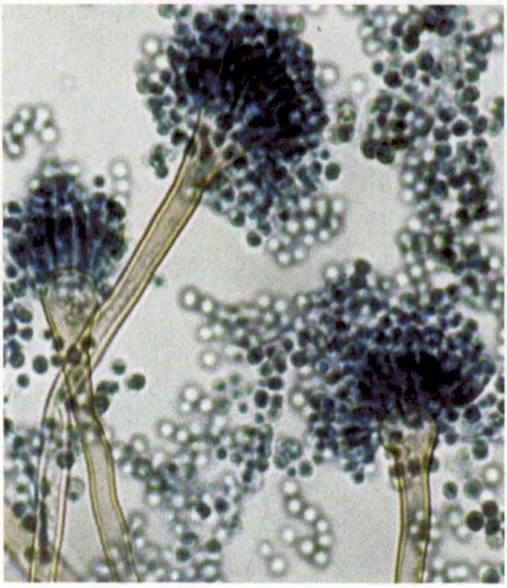

Fig. 10.21: Structure of *Aspergillus* sp.

10.5.2. Fusarium

The hyphae of the fungus are hyaline, branched, and septate. Conidia are formed in slimy, effuse sporodochia called pionnotes, or sometimes scattered on the mycelium. The conidia, which are slimy, are of two types, i.e. micro-conidia (one-celled, oval or comma-shaped, pyriform or elongate) and macro-conidia (which are hyaline or pale spindle-shaped with pointed ends). They are two or more-celled, thick-walled, smooth, and cylindrical or sickle-shaped (Fig. 10.22).

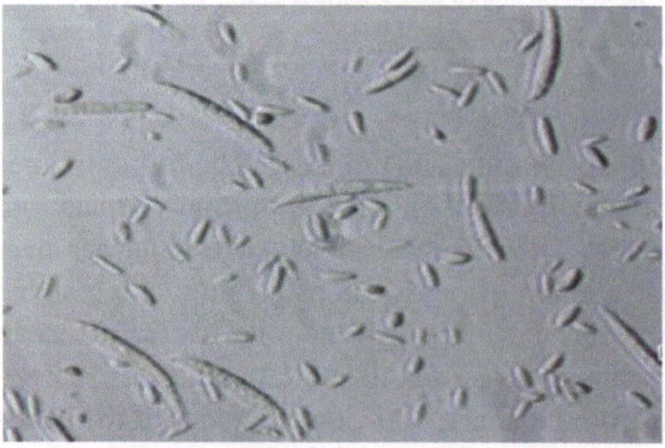

Fig. 10.22: Micro and macro-conidia of *Fusarium* sp.

10.5.3. Curvularia

The fungal colonies are pale brown to dark and fast growing. Mycelia are brownish, and cottony on the PDA. Conidiophores are multi-septate, straight to flexuous, simple, branched, brown, and bearing spores apically. Conidia are 3–5 celled, fusiform, clavate, 26–28 µm in length, 8–10 µm in width, and slightly curved at the third cell from the base. Basal and apical cells are pale or sub-hyaline, intermediate cells are brown, and the third cell from the base is darker (Fig. 10.23).

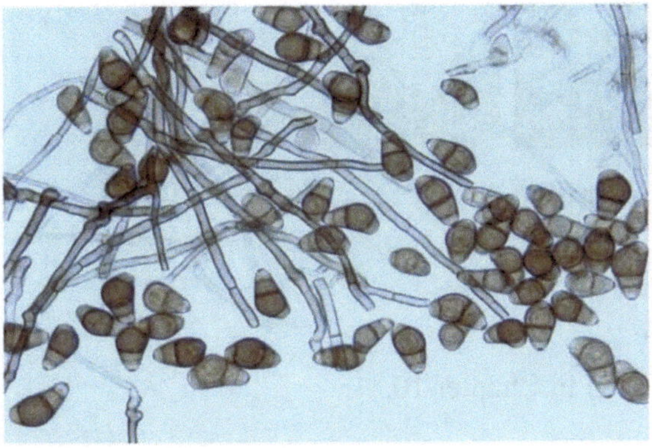

Fig. 10.23: Conidia of *Curvularia*.

10.5.4. Alternaria

Please refer "Alternaria" section in Chapter 6.

10.5.5. Phoma

The fungal colonies are grey to brown with red or pinkish pigmentation, delicate, fluffy, and concentrically zoned. Conidiophores are oblong, elliptic, or often irregular. The growth rate varies widely, depending on the species. Dark, globose to sub-globose to pyriform pycnidia with single or multiple ostioles occur on or immersed in the agar and may form singly or in aggregates. The hyaline conidiogenous cells lining the inner walls of the pycnidia produce their conidia by percurrent proliferation, which are then extruded in slimy masses. The variably shaped conidia are unicellular and hyaline. Alternarioid chlamydoconidia are seen in some species. Boerema et al. have recently provided an excellent revision and identification manual for this genus. Conidia are cylindrical to oval, allantoid, hyaline, pale yellow to brown, and usually one-celled (Fig. 10.24).

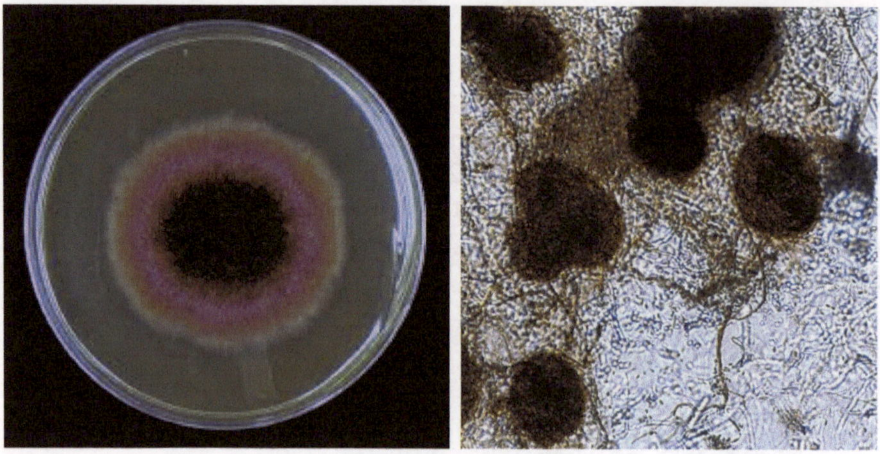

Fig. 10.24: Mycelial colony and pycnidia of *Phoma.*

10.5.6. Penicillium

Please refer "Penicillium" section in Chapter 10.

11

Identification of Fungi Causing Post-Harvest Diseases in Transit and Storage

11.1. Fruit Rots

11.1.1. Alternaria

Please refer "Alternaria" section in Chapter 10.

11.1.2. Colletotrichum

Please refer "Colletotrichum" section in Chapter 9.

11.1.3. Fusarium

Please refer "Fusarium" section in Chapter 10.

11.1.4. Penicillium

Please refer "Penicillium" section in Chapter 10.

11.1.5. Rhizopus

Please refer "Rhizopus" section in Chapter 9.

11.1.6. Botrytis

Please refer "Botrytis" section in Chapter 9.

11.1.7. Aspergillus

Please refer "Aspergillus" section in Chapter 9.

11.1.8. Geotrichum

Geotrichum fruit rot, also known as sour rot is common in citrus fruits, tomatoes, and strawberries. It is caused by the fungus *Geotrichum candidum*.

The affected fruits have a water-soaked, watery soft rot with a sour smell, and sometimes with white mycelium on the fruit surface. The genus *Geotrichum* includes over 100 species, the most common ones are *G. candidum*, *G. clavatum*, and *G. fici*, all of which have pathogenic potential. Species of *Geotrichum* can be differentiated by sequence analysis of the nuclear ribosomal large subunit (18S ribosomal RNA) or the internal transcribed spacer region of the nuclear ribosomal RNA gene.

On the growth medium, *G. candidum* colonies are thin, spreading, soft, creamy, and white in the anamorph state. The fungus *G. candidum* is characterized by hyphae that appear creeping, mostly submerged and septate. The hyphae are hyaline or lightly pigmented. Conidia appear arthrosporous, terminal, or intercalary, aerial on an agar surface. The conidia size ranges from 4.8 to 12.5 µm × 2.4–2.5 µm. Blastoconidia sometimes develops on hyphae laterally. When the hyphae become airborne it changes shape from arthroconidia to cylindrical or barrel-shaped or ellipsoidal. Chlamydospores are subglobose, solitary, borne on undifferentiated hyphae.

Geotrichum candidum is thought to be homothallic but most isolates are self-sterile. Sexual reproduction was first observed in strains isolated from soils in Puerto Rico. The fungus produces globose asci that contain a single, thick-walled, uninucleated, globose to oval ascospore measuring 6–7 µm by 7–10 µm. The ascospores have a smooth inner wall and a furrowed outer wall. The septa are perforated by microspores, arranged in a ring structure. The colonies appear to be growing faster in the sexual stage than the asexual stage. Colonies grow at a rate of 5–7 mm daily at 24° C (Fig. 11.1).

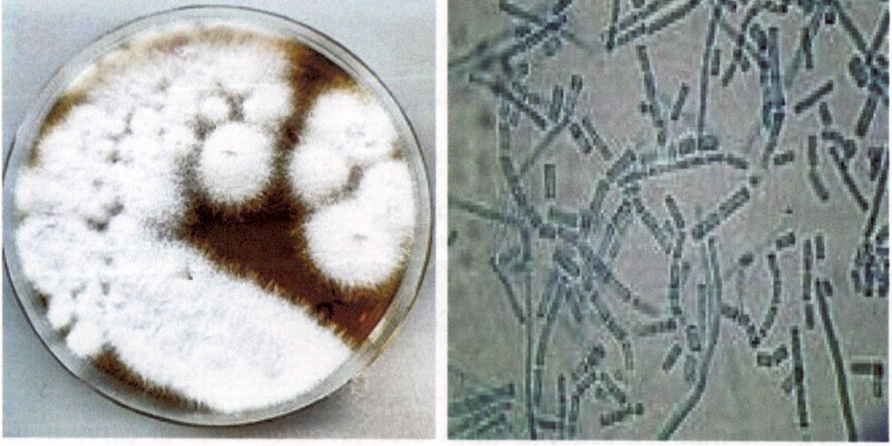

Fig. 11.1: *Geotrichum candidum* on malt-dextrose agar and microscopic fungal structure.

11.1.9. Sclerotinia

Please refer "Sclerotinia" section in Chapter 7.

11.1.10. Ascochyta

Please refer "Ascochyta" section in Chapter 6.

11.1.11. Botryodiplodia

Please refer "Botryodiplodia" section in Chapter 10.

11.1.12. Dothiorella

Dothiorella is a genus of fungi in the family *Botryosphaeriaceae*. There are three subgenera and approximately 322 species.

Dothiorella is characterized by pigmented, single septate ascospores and conidia that become brown and one septate while attached to the conidiogenous cells. Members of this genus are pathogens, endophytes, and saprobes in a wide range of hosts. Conidiogenous cells are cylindrical to fusiform, hyaline, thin-walled, smooth, and give rise to periclinal thickenings. Dark, oval to ovoid, often pigmented ellipsoid conidia (*Diplodia-like*) are hyaline, later brown, thick-walled, single septate before release from conidiogenous cells, occasionally slightly constricted at the septum with a broadly rounded apex and truncate base (Fig. 11.2).

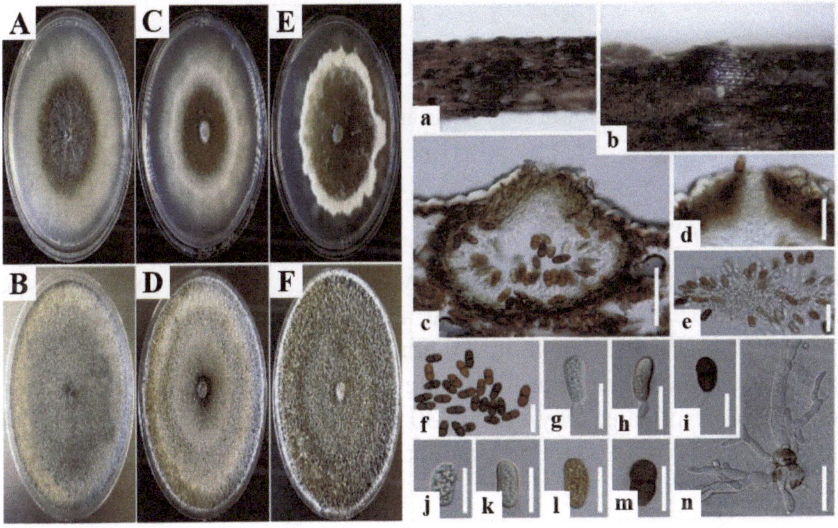

Fig. 11.2: *Dothiorella* mycelial growth **(A–F)**, pycnidia and spores (a-n) on conidiogenous cells.

11.1.13. Monilinia

There are three main fungal species: *Monilinia laxa, M. fructigena* and *M. fructicola*. The pathogen infects aerial parts of host plants with a variety of symptoms; it is also present as post-harvest pathogen. Conidiophores are branched and form chains of oval conidia, which are easily separated and dispersed by wind. The hyaline, ellipsoid, ovoid, lemon-shaped, and single-celled conidia are produced in chains in gray to light tan, cushion-shaped sporodochia on infected tissue. Microconidia (2.5–3 µm) may also be associated with *M. fructicola*; these are produced on bottle-shaped phialides on decayed fruit or in old cultures. Colonies on PDA are gray to brown with apparent mycelium tufts of sporulation (Fig. 11.3).

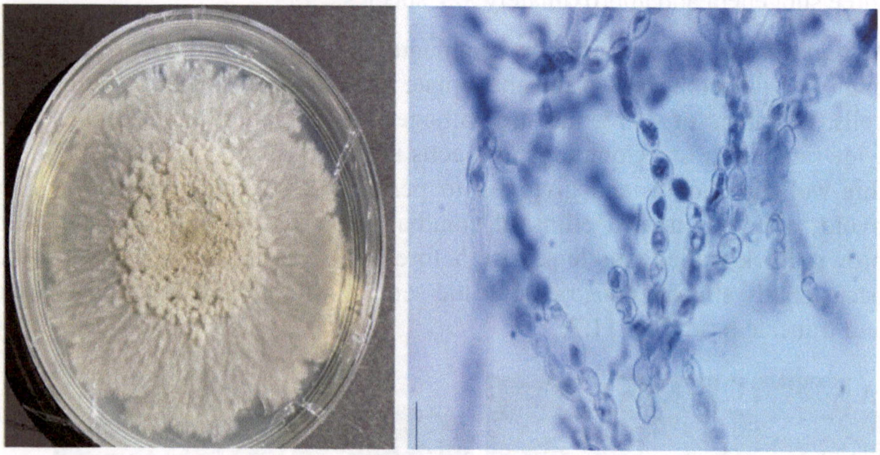

Fig. 11.3: Monilinia mycelial growth and conidia in chain.

11.1.14. Phomopsis

Microscopic observation of this genus reveals that it produces only one type of conidia, which are hyaline, one-celled, sub-cylindrical, and 5–9 × 2–2.8 µm in size, which gradually change into the beta form. Another form of conidia, the stylospores (β-conidia), are filiform, curved, hyaline, and aseptate. These spores do not germinate; they are 20–30 × 0.5–1.0 µm in size. In the PDA medium colony, color is white with a pinkish tinct (Fig. 11.4).

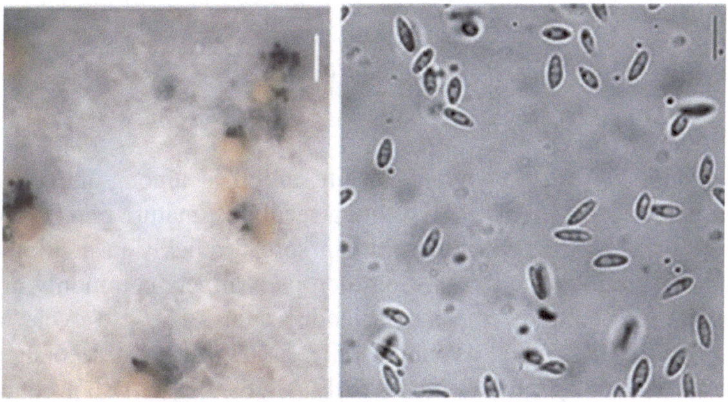

Fig. 11.4: Fungal growth and Conidia of Phomopsis.

11.1.15. Phytophthora

Please refer "Phytophthora" section in Chapter 6.

11.1.16. Thielaviopsis

Conidiophores of this genus are unbranched, straight, hyaline, smooth, septate, 30 × 6–8 μm, with the uppermost cell functioning as a conidiogenous cell and forming conidia. Conidia are unicellular, thick-walled, brown in chains of 4–6, remain intact for a long time, and resemble multi-septate conidia, terminal conidium thick-walled, brown, apex obtuse, and base truncate, while the rest of the conidia are longer than the wider thick-walled brown apex and base truncate 7–10 × 10–15 μm (Fig. 11.5).

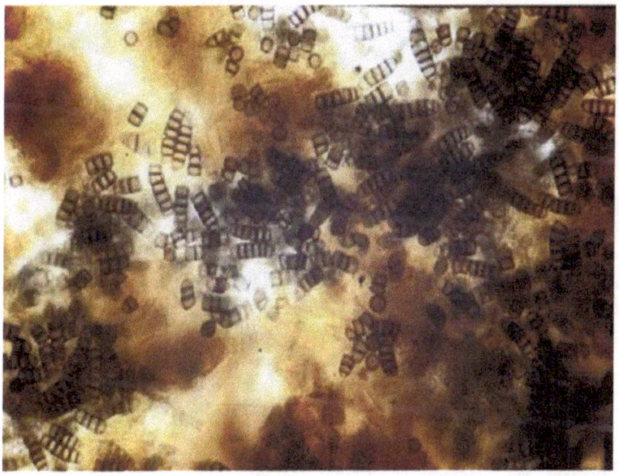

Fig. 11.5: Conidia of *Thielaviopsis*.

11.1.17. Verticillium

Please refer "Verticillium" section in Chapter 9.

11.1.18. Whetzelinia (Sclerotinia)

Sclerotinia sclerotiorum, also known as white mold, is among the most omnivorous of plant pathogens. It affects young seedlings, mature plants, and fruit in the field or in storage. White mold can spread quickly in a storage facility throughout the harvested crop. Some crops it affects commonly are soybeans, green beans, sunflowers, canola, and peanuts. The life cycle of *S. sclerotiorum* can be described as monocyclic, as there are no secondary inoculums produced. During late summer to early fall, the fungus will produce a survival structure called a sclerotium either on or inside the tissues of a host plant. *S. sclerotiorum* sclerotia can remain viable for at least three years and germinate to produce fruiting bodies called apothecia, which are small, thin stalks ending with a cup-like structure about 3–6 mm in diameter. The cup of the apothecium is lined with asci, in which the ascospores are contained. When the ascospores are released from the asci, they are carried by the wind until they land on a suitable host. The ascospores of *S. sclerotiorum* infect aboveground plant host tissue and begin to invade the host's tissues *via* mycelium, causing infection. *S. sclerotiorum* is capable of invading nearly all tissue types including stems, foliage, flowers, fruits, and roots. Eventually, white, fluffy mycelium will begin to grow on the surface of the infected tissues. At the end of the growing season, *S. sclerotiorum* will once again produce sclerotia. The sclerotia will then remain on the surface of the ground or in the soil, on either living or dead plant parts until the next season. The life cycle will then continue, respectively (Fig. 11.6).

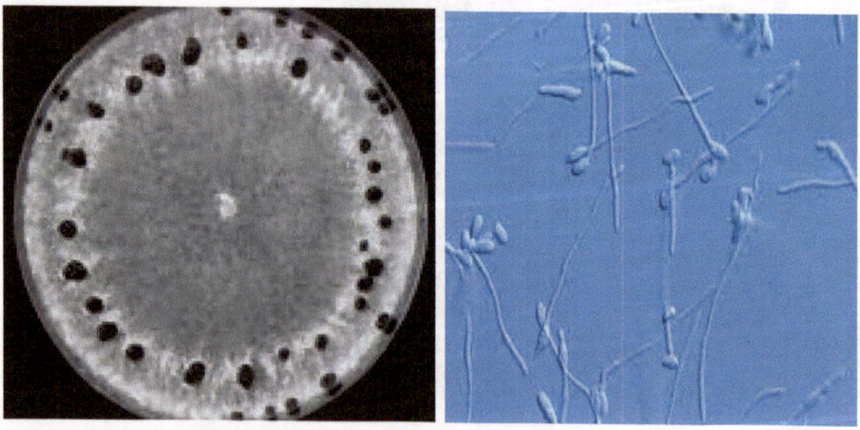

Fig. 11.6: *Whetzelinia* sclerotial bodies, and germinating ascospore.

11.2. Vegetable Rots

11.2.1. Botrytis

Please refer "Botrytis" section in Chapter 9.

11.2.2. Aspergillus

Please refer "Aspergillus" section in Chapter 9.

11.2.3. Cladosporium

Please refer "Cladosporium" section in Chapter 6.

11.2.4. Fusarium

Please refer "Fusarium" section in Chapter 10.

11.2.5. Penicillium

Please refer "Penicillium" section in Chapter 9.

11.2.6. Phytophthora

Please refer "Phytophthora" section in Chapter 6.

11.2.7. Phomopsis

Please refer "Phomopsis" section in Chapter 9.

11.2.8. Cercospora

Please refer "Cercospora" section in Chapter 6.

11.2.9. Rhizopus

Please refer "Rhizopus" section in Chapter 9.

11.2.10. Sclerotium

Please refer "Sclerotium" section in Chapter 5.

11.2.11. Sclerotinia

Please refer "Sclerotinia" section in Chapter 7.

12

Isolation of Fungal Pathogen for Culturing and Identification

12.1. Routine Media for Fungal Isolation

In recent years commercially available readymade culture media have been used in research laboratories for the isolation of fungal plant pathogens. These culture media are standardized and they are in different types. Some media are used for general purposes while some others are used for the isolation and growth of specific pathogens e.g. potato dextrose agar (PDA) is commonly used for the isolation and growth of a wide range of fungi in laboratories and its compositions are well-defined while certain specialized media are used for isolation of specific fungi like *Pythium*. Based on chemical composition, media can be classified into:

(A) Natural, (B) semi-synthetic, and (C) synthetic.

A. Natural medium

The exact chemical composition of this media is not known properly. It includes ingredients of natural origin like yeast extract, beef, milk, tomato juice, vegetable juices, blood, etc. Sometimes this medium is also referred to as a complex medium or non-synthetic medium because the medium is of complex type and contains various natural ingredients of unknown chemical composition. This type of media is useful for the cultivation of microbes whose specific growth factor requirement is not known. e.g. carrot slices, potato plugs, twigs, milk, meat extract, peptone, etc (Table 12.1).

Table 12.1: Some common useful natural media are described in the table.

Name of culture media	Composition	Preparation method
Tapioca potato sugar agar media (TPSA)	Potato-200 g Tapioca-20 g Sugar-20 g Agar-10 g	Boil 200 g of peeled potato with distilled water and sieve through a muslin cloth. In boiled potato extract, add 2% tapioca powder, 1% agar powder, and 2% sugar. Boil this mixture till homogenously mixed, sieved through muslin cloth, pour in appropriate quantities in flasks, and sterilize in autoclave. Use for a wide range of fungal pathogens.
Oatmeal agar	Rolled oats-30 g Agar-30 g Distilled water-1000 ml	Cook oatmeal in water for 15–30 min. Filter through three or four layers of cheesecloth and bring the filtrate back to volume with water. Add agar and autoclave to sterilize it.
Wheat meal agar (WMA)	Wheat-30 g Agar-30 g Distilled water-1000 ml	Add 30 g wheat flour and 30 g agar to 1000 ml distilled water, dissolve by boiling, and autoclave to sterilize it. Use for a wide range of fungi.
Potato carrot agar (PCA)	Potatoes-20 g Carrots-20 g Agar-15 g Distilled water-1000 ml	Dice potatoes and carrots and cook in water for 1/2 h. Strain through cheesecloth and add agar to filtrate. Autoclave at 121° C for 15 min. Good for some anamorphic fungi.
Carrot agar	Agar-10 g Fresh carrot-800 g Distilled water-1000 ml	Wash, peel, and dice carrots. Place carrots into a flask with 500 ml of water. Autoclave for 20 min. Blend the carrot in a blender until the mixture appears smooth. Add another 500 ml of water. Add agar and autoclave for 30 min.
Lima bean agar	Lima bean infusion from 62.5 g. Agar-15 g Distilled water-1000 ml	Suspend beans in 1000 ml of distilled water. Heat to boiling till it gets soft completely and crush. Sieve the infusion through a muslin cloth. Add agar to it and make the volume to 1 l. Sterilize by autoclave at 15 lbs pressure for 15 min. Mix well and pour in sterilized petri plates. The medium is used for phytopathological and other fungi.

Name of culture media	Composition	Preparation method
Rice extract agar	white rice-40 g Agar-20 g	Suspend 40 g of rice in 1000 ml distilled water. Heat to boiling to dissolve it completely, and add 10 ml polysorbate-80 and agar. Sterilize by autoclaving at 15 lbs pressure for 15 min. Mix well and pour into sterile Petri plates. Recommended for identification of *Candida albicans* using chlamydospore production.
Bread crumb agar	Bread crumbs, plain, unflavored 50 g Agar 9 g Distilled water 500 ml	Suspend 50 g bread crumbs with 9 g agar in 500 ml of distilled water. Boil it to dissolve it completely. Mixing well, dispense as required, and sterilize. This media is mainly used for *Armillariella mellea.*
Rabbit dung agar	Fresh rabbit dung-500 g	Place rabbit dung in a dry flask. Autoclave for 30 min. Pour the required quantity into a petri dish, and add a thick layer of sterilized 2% agar. Used for *Pilobolus crystallinus* and *Coprinus cinereus*
Malt agar (MA)	Malt extract 30 g Agar 15 g Distilled water 1000 ml	Add 30 g malt extract and 15 g agar to 1000 ml distilled water, boil, and autoclave to sterilize it. Used for a wide range of fungi. It is useful for culturing many Ascomycota. An all-purpose type of medium.

B. Semi-synthetic

The chemical composition of media is only partially known. Some common semi synthetic media are listed below:

Name of culture media	Composition	Preparation method
Water agar (WA)-	Agar-20 g Water-1000 ml	Add 20 g agar to 1000 ml water, boil and autoclave it. Used for isolating fungi from surface-sterilized substrates.
Potato dextrose agar (PDA)	Potatoes-200 g Dextrose-20 g Agar-20 g Water-1000 ml	Cut 200 g of peeled potatoes into small pieces, suspend in 1000 ml of distilled water and boil for 30 min. Decant the extract or filter through muslin cloth and make the final volume to 1000 ml. Add dextrose and agar into it, boil for 20 min and autoclave it. It is a relatively rich medium for growing a wide range of fungi.

Name of culture media	Composition	Preparation method
Malt extract agar (MEA)	Malt extract 30 g Peptone 5 g Agar 15 g Distilled water 1 l	Dissolve agar, malt extract, peptone in water, and boil, and mix it well. Dispense as required and sterilize in autoclave. A good growth medium for soil fungi, fungi isolated from wood, basidiomycetes, etc.
Corn meal agar (CMA)	Corn meal 20 g Peptone 20 g Dextrose 20 g Agar 15 g Distilled water 1 l	Suspend all the ingredient in 1 l of distilled water, and heat to boiling to dissolve the ingredients. Add 1% polyxyethylene-sorbitan-monooleate (Tween 80). Sterilize the medium in autoclave at 15 psi for 30 min. Use as per the requirement protocol. A good growth medium for anamorphic fungi, fungal sporulation and chlamydospore production, etc.
Hemp seed agar	Hempseed-50 g Agar-17 g Distilled water-1000 ml	Add the hempseed in sufficient quantity of water and soak overnight. After soaking autoclave it for 15 min, strain broth through cheese cloth. Mix agar and bring volume with distilled water to 1 l. Autoclave for 15 min before pouring in plates/test-tubes. To induce sexual reproduction of *Phytopthora capsici* and *Phytopthora parasitica*
Apple juice agar	Apple juice concentrate-300 ml Sucrose-10 g Agar-20 g Distilled water-700 ml	Suspend all ingredients in 700 ml of distilled water. Bring to the boil to dissolve the ingredients completely. Mix well. Dispense as required and sterilize.

Name of culture media	Composition	Preparation method
Vegetable juice agar (V-8)	V-8 juice-100 ml L-Asparagine-10 g Yeast extract- 2 g Calcium carbonate-2 g Glucose-2 g Agar-20 g Distilled water-1000 ml	Suspend all ingredients in 1000 ml of distilled water. Heat just to boiling. Sterilize by autoclaving at 15 lbs pressure (121° C) for 15 min. If slight precipitate appears after sterilization mix evenly before dispensing. Note: Due to presence of calcium carbonate, the prepared medium forms opalescent solution with white precipitate. Recommended for cultivation of yeast and molds.

C. Synthetic medium

The chemical composition of the medium is completely known. These media are very useful in studying the physiology, metabolic nature, and nutritional requirements of microbes. Both autotrophs and heterotrophs can be grown in these media.

Name of culture media	Composition	Preparation method
Raulins medium	Glucose-50 g Tartaric acid-3 g Ammonium tartrate-3 g Ammonium hydrogen Phosphate-0.4 g Ammonium sulphate-0.2 g Potassium carbonate-0.4 g Magnesium carbonate-0.3 g Iron (II) sulphate-0.05 g Zinc sulphate-0.05 g Distilled water 1 l	Dissolve the ingredients in water by boiling it and the medium may be sterilized by steaming for 30 min on each consecutive day for three days.

Name of culture media	Composition	Preparation method
Richard's synthetic agar media	Potassium nitrate-10 g Monopotassium dihydrogen phosphate-5 g Magnesium sulphate-2.5 g Ferric chloride-0.02 g Sucrose-50 g Agar-15 g Distilled water 1 l	Dissolve all the ingredients in 1000 ml distilled water. Heat to boiling to dissolve the ingredients completely. Sterilize by autoclaving at 15 lbs pressure (121° C) for 15 min. Mix well and pour into sterile Petri plates. Richards synthetic agar is used for isolation and cultivation of fungi from soil samples.
Brain-heart infusion (BHI) agar	Pancreatic digest of casein-16 g Brain heart infusion from solids-10 g Peptic digest of animal tissue-5 g Sodium chloride-3 g Disodium phosphate-2.5 g Dextrose-2 g Agar-13.5 g	Suspend all ingredients to 1000 ml of distilled water and sterilize the media. It is a non-selective fungal culture medium that permits the growth of variety of microorganism.
Sabouraud's dextrose agar (SDA)	Peptone-10 g Glucose-20 g Agar-20 g Distilled water-1000 ml	Suspend all ingredients to 1000 ml of distilled water, boil and sterilize the media. Sabouraud's dextrose agar (2%) is most useful as a medium for the subculture of fungi recovered on enriched medium to enhance typical sporulation and provide the more characteristic colony morphology.

Name of culture media	Composition	Preparation method
Czapek's agar	Sucrose-30 g Sodium nitrate-2 g Dipotassium phosphate-1 g Magnesium sulphate-0.5 g Potassium chloride-0.5 g Ferrous sulphate-0.01 g Agar-15 g Distilled water-1 l	Dissolve the ingredients in 1000 ml distilled water, boil, and autoclave it. Recommended for the general cultivation of fungi from water samples.

Based on consistency the media are of three types as (A) liquid media, (B) semi-solid media, and (C) solid media:

A. Liquid media

Nutrient broth is the common liquid media used in plant pathological laboratory. It is routinely used for the isolation of fungal metabolites, and to study the growth mass of the fungal pathogens.

B. Semi-solid media

A semisolid medium is prepared with agar of concentration of less than 10%, and is useful in studying the rhizoid structures of the fungus.

C. Solid media

If agar is added to a broth, it becomes solid media. It is used for isolating fungal plant pathogens and to study the characteristics of the isolated fungal colonies. It remains solid on incubation and not destroyed by proteolytic bacteria.

12.2. Special Media for Isolation of Specific Fungal Pathogens

It is possible to isolate specific fungal plant pathogens by using selective special media irrespective to the category of the medium. Suitable selective media has to be identified for different pathogens (Table 12.2). Some selective special media are given below:

Table 12.2: Selective media.

Name of culture media	Composition	Preparation method
Oat flake agar	Oat flakes-30 g Agar-15 g Distilled water-1 l	Boil oat flakes in water for 10 min; fill up to 1 l and add agar. Autoclave 20 min at 121° C. Shake before pouring. pH 6.0. Oat flake media mainly used for *Physarum polycephalum*.
PCNB medium	Difco peptone-15 g K_2HPO_4-1 g $MgSO_4 \cdot 7H_2O$-0.5 g Agar-15 g PCNB-1.0 g Streptomycin-0.3 g Neomycin sulphate-0.12 g	Suspend all ingredients in 1000 ml distilled water. Boil to dissolve all ingredients completely. Mix well. Dispense as required and sterilize. This medium is useful for quantitative estimates of propagules of *Fusarium solani* from soil.
Apple juice agar	Apple juice concentrate-300 ml Sucrose-10 g Agar-20 g Distilled water-700 ml	Suspend all ingredients in 700 ml distilled water. Boil to dissolve the ingredients completely. Mix well. Dispense as required and sterilize. Used for *Venturia inequalis.*
Rose bengal agar	Dextrose-10 g Soy peptone-5 g Monopotassium phosphate-1 g Magnesium sulphate-0.5 g Chloramphenicol-0.1 g Rose bengal-0.05 g Agar-15 g Distilled water-1000 ml	Suspend all ingredients in 1000 ml distilled water. Boil to dissolve the ingredients completely. Mix well. Dispense as required and sterilize. Rose bengal agar with chloramphenicol is recommended for use in qualitative procedures for selective cultivation, isolation, and enumeration of yeast and molds from environmental specimens and food. Smith and Dawson added rose bengal to a medium with a neutral pH to selectively isolate fungi from soil samples.
Potato dextrose-yeast extract agar (PDYA)	Potato-250 g Agar-20 g Dextrose-10g Yeast extract-1.5 g Distilled water-1 l	Prepare the media like PDA and autoclave it. This media is good for growing cultures derived from mushrooms.

Name of culture media	Composition	Preparation method
Honey peptone agar	Pure honey-60 g Peptone-10 g Agar-20 g Distilled water-1 l	Suspend all ingredients in 1 l distilled water. Heat to dissolve. Dispense as required and sterilize. Media particularly used for yeast culture and *Eurotium chavalieri*
Malt glucose yeast peptone (MGYP) agar medium	Malt-extract-3 g Glucose-10 g Yeast extract-3 g Peptone-5 g Distilled water-1 l Agar-20 g Adjust pH to 6.4–6.8	Suspend all ingredients in 1 l of distilled water. Heat to dissolve. Dispense as required and sterilize. Media is particularly used for yeast culture.
Mannitol yeast extract agar	Agar-10 g K_2HPO_4-0.5 g $MgSO_4 \cdot 7H_2O$-0.2 g NaCl-0.2 g $CaCl_2 \cdot 6H_2O$-0.2g Mannitol-10 g Yeast extract-0.4 g distilled water- 1 l	Suspend all ingredients in 1 l of distilled water. Heat to dissolve. Dispense as required and sterilize. Media is particularly used for yeast culture.
Sordaria culture agar	Cornmeal agar-17.0 g Dextrose-2.0 g Yeast extract-1.0 g Distilled water-1.0 l	Suspend all ingredients in 1 l of distilled water. Heat to dissolve. Dispense as required and sterilize. This media is mainly used for *Sordaria fimicola.*
Rapeseed agar	Rapessed-100 g Agar-20 g Distilled water-1 L	Wash rapeseed thoroughly in water and boil in 1 l distilled water for 30 min. Filter extract through muslin cloth, add agar and adjust volume to 1 l, boil to homogenize and autoclave it. Suitable for production of oospores of *Phytopthora* spp.

Name of culture media	Composition	Preparation method
Pythium selective medium (P5ARP)	Part I. Corn meal agar-17 g; Water-1 l Part II Primaricin-0.2 ml Ampicillin-0.25 g Rifampicin (10 mg/l)-100 µl PCNB (100 mg A.I. = 142.9 mg 70 WP) − 0.143 g	1. Mix ingredients from part 1 and autoclave. 2. Cool to 50° C and add ingredients from Part 2. Notes Use a sterile syringe for Primaracin. Rifampicin (10 mg/100 µl dissolved in DMSO). For Phytopthora selective media add 50 mg ai hymexazol = 71.4 mg 70WP

Name of culture media	Composition	Preparation method
Phytophthora selective medium (P10VP + H) (For selective isolation of many species of *Phytophthora* from soil or roots).	Part I. Corn meal agar, 17 g; Water, 1.090 L Dextrose, 2g. Part II PCNB (pentachloronitrobenzene) solution (12.5 mg/ ml), 11.7 ml; Pimaricin solution (25 ml/ml), 880 µl (22 mg); Vancomycin, 220 mg; Hymexazole (Tashigaren), 72mg.	1. Mix ingredients from part 1. 2. Place stir bar in flask before autoclaving. Autoclave media and cool to 50–55° C. 3. While media cools heat PCNB solution in 65° C in water-bath to dissolve. 4. Add the ingredients from part 2 to cooled media. Mix Note: To prevent bacterial contamination during culturing, it is essential that plant tissues and the agar surfaces are dry. Blot plant tissue with clean paper towels. P10VP is excellent for the selective isolation of various species of *Phytophthora* from plant tissues and soil. The selectivity is enhanced by the addition of hymexazole, which inhibits the growth of *Pythium* and *Mortierella.* However, some species of *Pythium* will grow on this medium and could be misidentified as *Phytophthora.* Hymexazole at 50 mg/l is found to inhibit radial growth of *Phytopthora cactorum* and *Phytopthora palmivora* by 50–80%.

Name of culture media	Composition	Preparation method
		Pimaricin is a broad-spectrum antifungal antibiotic that inhibits the true fungi but not members of the *Pythiaceae*. Once in solution, it is inactivated by light so that culture plates must be stored and incubated in the dark. Growth of septate fungi indicates the inactivation of pimaricin. Concentrations above 10 mg a.i./l inhibit oospore germination of *Pythium* and *Phytophthora*. Vancomycin is a broad-spectrum antibiotic active against both Gram negative and Gram-positive bacteria. It can be used as a substitute for penicillin and polymyxin. Vancomycin can be inhibitory to some species of *Phytophthora* at concentrations exceeding 400 mg/l. PCNB has a relatively narrow spectrum of activity against fungi. Hyphal morphology of many *Phytophthora* species is distinct on corn- meal based agar.
Selective medium for *Rhizoctonia solani* (Ko and Hora 1971)	KH_2PO_4-1.0 g $MgSO_4 \cdot 7H_2$-0.5 g KCl-0.5 g $FeSO_4 \cdot 7H_2O$-0. 01 g $NaNO_2$-0.20 g Gallic acid-0. 40 g Agar-20.0g Distilled water-1 l	Suspend all ingredients in 1 l of distilled water. Heat to dissolve. Dispense as required and sterilize. Note: For isolation from seeds or soil add Dexon, 90 mg (a fungicide to suppress *Pythiaceae* fungi); Chloramphenicol, 50 mg; and Streptomycin, 50 mg (to suppress bacterial growth) in 1 l of the medium.

Name of culture media	Composition	Preparation method
Lactose-mineral-salts medium (Frank and Somkuti 1979)	Lactose-10.0 g KH_2PO_4-5.0 g NH_4Cl-2.0 g $MgSO_4 \cdot 7H_2O$-0.25 g NaCl-1.0 g distilled water-1 l pH-7.0	Suspend all ingredients in 1 l of distilled water. Heat to dissolve. Dispense as required and sterilize. Media is particularly used for yeast culture.

12.3. Isolation of Fungi from Infected Sample

i. Select infected host tissue from the advancing margin of the lesions.

ii. Cut into small pieces (2–5 mm) containing both the diseased and healthy tissue and keep in sterile Petri dishes.

iii. Dip the pieces into 1% sodium hypochlorite or mercuric chloride solution for about 1 min.

iv. Transfer the pieces to Petri-dishes containing sterile distilled water and wash thoroughly in two changes of sterile water to free them from the chemicals, if any.

v. Rub hands with rectified spirit and wipe the table top of inoculation chamber with rectified spirit.

vi. Lit the burner

vii. Hold the flask containing sterile luke warm PDA in the right hand and remove plug near the flame. Lift the lid of Petri dish gently with left hand and pour about 20 ml of medium. Close the mouth of the flask with plug near the flame.

viii. After solidification of the medium, place four sterilized pieces of diseased portions at different distance in a single PDA plate.

ix. Incubate the Petri dishes in an inverted position at 25° C and examine for 3-5 days and observe the incubated plates.

12.4. Purification of Fungal Culture

12.4.1. By Hyphal Tip Method

In this method the growing hyphal tip is selected and transferred in to the media slant to obtain the pure culture.

i. Select the single growing hyphal tip of the mycelium with the help of microscope and mark it with glass marking pencil.

ii. Cut the marked portion with the help of sterile cork-borer and transfer it on the PDA slant and incubate at the temp of 27 ± 2° C and observe for pure growth of culture.

12.4.2. By Single Spore Isolation

i. Prepare the spore suspension from the sporulating fungus in order to have 10^6 spores/ml.

ii. Add above (10 ml) spore suspension in 250 ml of liquefied luck-warm (when temperature of medium is around 40° C) potato dextrose agar medium.

iii. Add the medium in sterilize Petridish and allow to solidify it.

iv. After the solidification of the medium locate the single spore with the help of microscope and mark with a glass marking pencil or immediately after appearance of fungal growth, locate the single well isolated colony and mark it.

v. Cut the marked portion with the help of sterile cork-borer and transfer it on the PDA slants.

vi. Incubate the slants for few days and observe for pure growth of culture.

12.5. Preservation and Maintenance of Fungal Culture

General or specialized cultural media is used for short or medium-term storage of fungal cultures. Maintenance and preservation of fungal cultures helps to maintain it in the original form. There are several methods used for preservation of these cultures.

12.5.1. By Periodic and Serial Transfer

This method is generally adopted for culture of non-sporulating fungi or for those having short spore lives. The fungal cultures are transferred periodically at regular interval on the media slants. Culture can be stored at 5–8° C up to 2–3 months.

12.5.2. By Storage in Distilled Water

Fragment of hyphae or spore can be transfer into new screw capped vials or tubes half filled with sterile distilled water and caps are tightened to restrict the evaporation of water and stored in refrigerator at room temperature. This is another inexpensive and low maintenance method for storing fungal cultures for six months to one year without getting contamination.

12.5.3. In Mineral Oil

Fungal cultures are covered with sterile mineral oil at least 1 cm above the culture slant to prevent dehydration of medium and to reduce metabolic activities and growth of culture. Some fungi which produce acid or liquefy the medium are not suitable for this preservation. e.g. *Fusarium*.

12.5.4. In Dried Form

Pathogen in dried infected host tissue or on agar media slant with the resting spores or fruiting bodies are stored in low humidity in refrigerator or at ambient temperature for a long time period.

12.5.5. By Low Temperature Preservation

Lots of fungus culture can be preserved at low temperature, in order to inactivate their metabolism for short and long period of time. The fungal cultures can be stored below the freezing point temperature.

12.5.6. By Lyophilization

Lyophilization, also known as freeze-drying, is a preservation method achieved through dehydration. With lyophilization, the fungal cultures are frozen and subsequently dried under vacuum. The method is highly successful with cultures that produce mitospores.

The freeze-drying process occurs in several steps. Initially organisms are cultured using good aseptic technique, preferably on nonselective media without the addition of antibiotics. Once grown, a cell suspension is made in a cryopreservation medium/buffer. The samples are then subjected to the freeze-drying process involving freezing at a temperature below −40° C and then controlled drying under vacuum. The freeze-dried strains are stored between 4 and 10° C and are tested regularly for viability.

12.5.7. By Freezing in Liquid Nitrogen

Fungal cultures are grown directly in firmly closed sterile plastic cryovials (1.8 ml) with 200 mg of perlite (Agroperlit, agricultural grade) moistened with 1 ml of wort (4° Balling) or other medium (e.g., MYA) enriched with 5% glycerol as a cryoprotectant. For sterilization of the cork borer, lancet, and small spoon use a hot-air sterilizer (150° C, 30 min).

i. Distribute perlite into cryovials (200 mg per vial), flood it with 1.8 ml of the medium enriched with 5% of glycerol, and sterilize vials in autoclave (121° C, 20 min). In a laminar flow box cut out an agar plug (6 mm diameter) from the actively growing part of a colony on a Petri dish using a cork borer, place it using a sterile lancet on the surface of perlite in the cryovial, close the vial firmly and let it incubate for 14 days at 24° C.

ii. Freeze the cryovials with perlite overgrown by mycelium in a programmable freezer (or alternatively in a "Mr. Frosty" container in a deep freezer) to −70° C at a freezing rate of 1° C per minute. Then place them in liquid nitrogen (LN) in a container.

iii. Take the stored frozen cultures in cryovials out of the LN container, transfer them to a water bath (37° C), and leave them there until the ice is completely thawed (thawing-reactivation of cultures). Prior to opening, disinfect the surface of cryovials with 70% ethanol.

Viability test: The fungal cultures with this storage method are reported to be stored for up to 13 years.

12.5.8. By Direct Freezing of Cultures

Fungal cultures grown on agar slants in bottles or test tubes with screw caps are placed directly in the freezer. These cultures are preserved successfully at −80° C for up to two years. In general, vigorously growing and sporulating cultures survive the freezing process better than less vigorous strains.

12.5.9. In Soil Preservation

Some fungi can be preserved easily and successfully for many years in dry, sterile soil or sand. This is low maintenance and cheap method very useful for soil fungi i.e. *Rhizoctonia* sp. Fungal culture suspension is poured into sterile soil, allowed to grow for 10–12 days and such fungal enriched soils can then be stored at ambient temperature.

12.6. Induction of Sporulation in Fungi

Spores and fruiting structures are the most important morphological characters used by mycologists to classify fungi into genera or differentiate closely related species. However, numerous isolates fail to sporulate on commonly used media. Fungal sporulation is a complex process influenced by environmental and endogenous biological rhythms.

The factors which can induce fungal sporulation includes nutritional, starvation media, use of plant tissues (leaves, stem, petioles, etc.), use of alternate light and dark cycles, use of slide culture technique, use of $CaCO_3$, and use of low-strength nutrient media etc.

12.6.1. Use of Nutritional Starvation

Many nutritional factors are effective in fungal sporulation, e.g., the carbon source, nitrogen source, and microelements. Some fungi have specific carbon and nitrogen requirements for sporulation. Potato dextrose agar (PDA) is the most widely used medium in fungal isolation and culture. Other commonly used media include potato carrot agar, malt dextrose agar, yeast extract-phosphate medium, Czapek yeast autolysate agar, malt extract agar, cornmeal agar, potato sucrose agar, V8 vegetable juice agar. These media provide favorable growth condition for most endophytic and pathogenic fungi, but are often ineffective in inducing sporulation of sterile isolates. Sporulation usually occurs when growth rate is reduced and is hampered under conditions that favor rapid mycelial growth. Starvation or nutritional depletion often stimulates sporulation, and some modifications of artificial media with low nutrient have been applied. Some typical low nutrient media include water agar media, half-, or 1/4-strength PDA, and synthetic nutrient-poor agar medium.

Heavy sporulation of *Colletotrichum gloeosporioides* from mango is observe if maltose and ammonium nitrate are used as carbon and nitrogen sources, respectively. Further *C. gloeosporioides* sporulates when the isolates is pre-cultured in potato dextrose broth and then spread on diluted OMA under continuous light. Low nutrient is helpful in inducing sporulation in *Colletotrichum* species.

12.6.2. Use of Slide Culture

Slide culture is often used in the induction of appressoria and conidiophores in *Colletotrichum*. Slide cultures are also a recommended method for conidial induction.

i. The slide is supported by two sticks above one or two layers of filter paper on the bottom of a petri dish, which are autoclaved altogether.

ii. The filter paper is soaked with sterilized water to maintain humidity.

iii. Two 0.5 mm × 0.5 mm pieces of PDA are placed on the slide and inoculate with a small quantity of mycelium at one corner of the agar block.
iv. Finally, a sterilized coverslip is placed on the agar block. The assembly is sealed and incubated at the optimum temperature for fungal growth for 7–14 days, after which the coverslip can be removed and observed microscopically.

12.6.3. Use of Host Tissue

Plant pathogenic fungi are often used to induce sporulation on the source host tissues. Some plant tissues are effective in inducing sporulation. Autoclaved leaves of *Rhododendron pulchrum* cv. ohmurasaki are used to induce sporulation in *Guignardia endophyllicola*. A sterilized petiole fragment of the palm *Livistona chinensis* is used to induce sporulation of endophytic sterile mycelia isolated from *L. chinensis*. Autoclaved pine needles are added onto the surface of agar media to induce pycnidia of some *Botryosphaeriaceae* spp. Sterilized branch segments of *Sophora japonica* are utilized to promote sporulation of the endophytic fungi from lichens. Carnation (*Dianthus caryophyllus*) leaves are found to be a good substrate for sporulation in *Fusarium* and *Pestalotiopsis* species. Leaves of *Hydrangea* or *Gardenia* are found to be effective for perithecia and ascospore formation of *Glomerella cingulata*, *Mycosphaerella allicina*, *Metasphaeria* sp., *Guignardia* sp., and *Pleospora herbarum*.

12.6.4. Use of Chemical Supplement

Some chemical supplements like $CaCO_3$ induced sporulation in some *Alternaria* species.

12.6.5. Use of Light Effect

Light has been regarded as an important stimulus for conidial formation. Exposure of mycelia to near-ultraviolet (UV) light and blue light has been widely used. For example, spores have been successfully induced by near-UV light irradiation in some species of *Ascomycetes*, *Basidiomycetes*, *Myxomycetes*, and *Zygomycetes*.

Wavelengths spanning 350–500 nm are most effective in inducing sporulation. Various fungal isolates showed different threshold dosages under particular wavelength, and excessive irradiation can inhibit sporulation. One commonly used method is 12 h near-UV light irradiation and irradiation in darkness for 12 h, imitating the diurnal cycles of light and darkness.

12.6.6. Other Factors

Though previous studies focused on factors that affect sporulation, such as temperature, pH, humidity, and mutation of the mycelium, these studies often focused on spore yield. Generally, combinations of these factors are often adopted to induce sporulation.

13

Testing Pathogenicity of Fungal Cultures (Koch's Postulates)

The pioneering German microbiologist, Robert Koch, identified a set of four conditions that need to be satisfied to establish that a particular organism is the causative agent of a particular disease. These conditions, known as Koch's postulates, are:

1. The suspected pathogen must be found associated with the disease in all the diseased plants examined.
2. The organism must be isolated from the diseased tissue and grown in pure culture on nutrient media, with its characteristics described (non-obligate parasites) or in a susceptible host plant (obligate parasites), and its appearance and effects recorded
3. When a healthy plant of the same species or variety is inoculated with this culture, it must produce the disease and show the characteristic symptoms.
4. The organism must be re-isolated from the inoculated plants and must be shown to be the same pathogen as the original.

If all the above steps have been followed and proved true, then the isolated pathogen is identified as the organism responsible for the disease.

13.1. To Prove the Pathogenicity of Seed-Borne Fungi

The method is used to demonstrate the pathogenic nature of seed-borne fungal pathogens and microflora. The seed-borne fungal pathogens can cause either seed decay, damping off, wilt, or symptoms on ground plant parts like leaf spots, blights, etc.

13.1.1. Seed Germination Test

The seeds are disinfected with 1% Clorox, rinsed thrice with distilled water, and treated with the culture of seed-borne fungal test pathogen. Such treated seeds are subjected to germination tests on germination papers in Petri plates or in sterile soil. The germinating seed can be observed for seed decay, radical rot, or other seedling symptoms and confirm Koch's postulate.

13.2. To Prove the Pathogenicity of Soil-Borne Fungi

Various methods of pathogenicity tests of soil and seed-borne fungi are listed below:

13.2.1. Seeding with Inoculum Disk Method

Seeds disinfected with 1% Clorox are sown in the autoclaved soil media (1:1:1 soil, sand, and peat moss) in the plastic pots. Three seeds are sown in a pot with a 5 mm disk of 7-day fungal culture (approximately containing 3.2×10^4 spores/disk). Disks are placed adjacent to seeds. The pots are incubated at 22–25 °C in the growth room by placing them on the racks randomly. After seven days the germinating seeds or seedlings are observed for damping off, or seed rot, or wilting symptoms and confirm Koch's postulate.

13.2.2. Root Dip Method

The method is used to demonstrate the pathogenicity of wilt and root rot pathogens. Seeds are sown in autoclaved sand and watered daily. After 6–8 days of the sowing, the raised seedlings are removed from the sand very carefully so that the roots remain intact. These roots of seedlings (injured or uninjured) are dipped in the spore suspension of the pathogen (3.2×10^4 spores/ml) and planted in pots already filled with autoclaved soil containing silt loam, sand, and peat moss (1:1:1). The pots are placed in growth room at 22–25 °C and the root rot, wilt, and foot rot symptoms are observed after four weeks and confirm Koch's postulate.

13.2.3. Stem Inoculation

The method is used to demonstrate the pathogenicity of foot rot pathogens. Pierce the lower stem at the ground level of the treatment plant with a sterile inoculating needle or hypodermic needle and place a small piece of agar from a pure culture of the pathogen onto the wound site (or inject a small volume of spore suspension into the stem using a hypodermic syringe and needle). Pierce the lower stem of the control plant with the sterile inoculating needle (or with the hypodermic needle), but do not treat with the inoculum. Wrap parafilm or plastic wrap over the wounds or injection sites or cover the growing treated plant with a polythene bag cover having aeration holes. Water the soil each day. Examine and compare the inoculated plants with the un-inoculated plants. Observe and record symptoms and compare these with symptoms observed in the field.

13.2.4. Seed Test Tube-Cotton Swab Method

The method is used to demonstrate the pathogenicity of seedling rot pathogens. Test tubes (20 cm × 3 cm) are prepared by filling 1/4th of cotton in the bottom of the tube. Sterilized distilled water in sufficient quantity is added to each tube to moisten the cotton and excess water is drained out. The test tube beak is covered with aluminum foil and is autoclaved. Test seeds (three seeds/tube) are surface disinfected with 1% Clorox and placed on the moist cotton swab in the test tube. One disk of 5 mm of fungal isolate (3.2×10^4 spores/disk) is placed adjacent to the seeds. After inoculation tubes are again covered with aluminum foil and are placed in a growth room at 25 °C to observe the seedling rot symptoms.

13.2.5. Sick Plot Methods

The method is used to demonstrate the pathogenicity of wilt, root rot, foot rot, and damping off pathogens. Sterilize the plot soil with the soil solarization method to make it free from unwanted pathogens. Now, prepare these plots of soil by incorporating bulk inoculum of the test pathogen into them. The process of incorporating inoculum should be continued till the soil is enriched completely (10^{10} cfu/g of soil) with the pathogen. Such plots are referred to as sick plots as the growth of a particular plant species is succumbing to the infection of the pathogen present in such sick soil. Whenever the crop (seed or seedlings) is sown, planted, or transplanted in such soils, it gets infected with the disease pathogen and killed. Isolate the pathogen from such a diseased plant and confirm with the added inoculum.

13.3. To Prove the Pathogenicity of Leaf Spot, Blight, and Blast Fungi

13.3.1. Detached Leaf Method

In detached leaf assay, the fourth to sixth sub-terminal leaves of test plants are detached from plants or plotted plants of 30 days old so that all leaflets are of similar age and size. These leaflets are washed in tap water to remove dirt and dust, appropriately surface sterilized with disinfectant, and then placed on moistened filter paper in glass petri plates. These leaves are inoculated with conidial suspension (containing 1×10^4 conidia/ml) of the test pathogen. A single leaflet is used as an experimental unit and 4–5 replications are conducted for each test. The leaflets inoculated petri plates are then placed in an incubator at 27 ± 1 °C and examined every 12 h still seven days of post-inoculation. During each observation, the severity of infection is rated by assessing the infected proportion of the leaflets with visible disease symptoms. The leaf sprayed with distilled water can serve as a control.

13.3.2. Spore Suspension Spray Method

The conidial suspension is prepared by flooding the fungal inoculum plates with sterilized distilled water, and the spore concentration is determined with a hemocytometer. A drop of Tween-20 is added per 100 ml of inoculum suspension before application. This spore suspension (containing approximately 3–4 × 10^4 spore/ml) is sprayed on the previously raised seedling in the pots (at the 2–3 leaf stage after seeding or planting). Approximately 15 ml of spray inoculum is required per plant with a sprayer. The plants are covered with plastic bags to attain high humidity for 30 h and then are placed in a growth room, glass-house, or net-house at 22–25 °C, note the disease severity, and confirm Koch's postulate.

13.4. To Prove the Pathogenicity of Die-Back Fungi

The pure culture of die-back fungal pathogen on the growth media in petri plates is collected by scraping it with a scalpel. Place this culture inoculum on the sterile moist cotton swab. Select the test plant host of the die-back pathogen and the growing apical tip portion of the stem or branch is clipped with scissors. The cotton swab with fungal inoculum is placed on the clipped tip of the stem and tied with a polythene cover. If such plants are in pots should be placed in a glass house or net house at 22–25 °C, note the die-back symptoms develop after 10–15 days of inoculation, and confirm Koch's postulate.

13.5. To Prove the Pathogenicity of Obligate Fungi

13.5.1. Rust Fungi

Rust fungi cannot be cultured on culture media commonly used in the laboratory. To prove the pathogenicity of the rust fungi, collect the rust inoculum from the infected source plant or obtain it in its pure form as lyophilized spore inoculum from a recognized laboratory. Make the spore suspension of the rust inoculum (the infected rust samples are immersed in distilled water; rub the rust pustules to release the uredospore and teliospores as rust inoculum) and inject into the leaf culm or spray on the leaves of the plant and stems. The sprayed plants should be covered with polythene bags to create the humidity needed for germination of rust spores and infection. Isolate the rust spores from newly developed rust pustules and compare them with the inoculated spore inoculum to prove Koch's postulate.

13.5.2. Smut and Bunt Fungi

Smut and bunt fungi cannot be cultured on culture media commonly used in the laboratory. To prove the pathogenicity of the smut and bunt fungi, collect the smut or bunt inoculum from the infected source plant or obtain it in pure form as lyophilized spore inoculum from a recognized laboratory. Make the spore suspension of the smut and bunt inoculum (the infected smut and bunt samples are immersed in distilled water; rub the smut pustules to release the teliospores as smut and bunt inoculum) and spray on the inflorescence of the plant. The sprayed inflorescence should be covered with polythene bags to create the humidity needed for the germination of teliospores and infection. Collect the seed formed in such inflorescence and use it in the next season to notice the development of smut and bunt symptoms in the developing inflorescence. Isolate the smut and bunt spores from newly developed smut and bunt and compare them with the previously inoculated spore inoculum to prove Koch's postulate.

13.5.3. Downy Mildew Fungi

The downy mildew infection can be from a soil source (as in the case of pearl-millet green ear disease) or an airborne source (as in the case of grapevine downy mildew). The downy mildew source inoculum can be zoospores or oospores.

For soil-borne infection of downy mildew, incorporate the oospore inoculum (in the form of chopped infected mildew material) in the sterile pot soil. Sow the seeds of the test plant in these pots, and cover the pots with moist gunny bags till the seedling emerges. Regular agronomic practices should be carried out. Observe the plants for the development of leaf or ear-head symptoms. Compare the isolated pathogen with the original incorporated inoculum to prove Koch's postulate.

For leaf spot downy mildews, collect the zoospore inoculum of downy mildew fungi in distilled water. Spray this inoculum on the test seedling or plant leaves and cover with a polythene bag to create the required humidity for spore germination and infection for at least 72 h. Keep the inoculated plants in the glasshouse or net house and observe for the development of downy mildew symptoms. Collect the inoculum of the pathogen from the newly developed symptoms and compare it with the original inoculum used for inoculation and disease development to prove Koch's postulate.

13.5.4. Powdery Mildew Fungi

For powdery mildews, collect the ascospore (from cleistothecial bodies) or *Oidium* inoculum of powdery mildew fungi in distilled water. Spray this inoculum on the test seedling and plant leaves and cover it with a polythene bag to create the required humidity for spore germination and infection for at least

48 h. Keep the inoculated plants in the glass- and net-house in a comparatively dry environment and observe for powdery mildew symptoms development. Collect the inoculum of the pathogen from the newly developed symptoms and compare it with the original inoculum used for inoculation and disease development to prove Koch's postulate.

13.6. To Prove the Pathogenicity of Inflorescence Fungi

The conidial suspension of the inflorescence fungal isolate is prepared by flooding the fungal inoculum plates with sterilized distilled water and the spore concentration is determined with a hemocytometer. A drop of Tween-20 is added per 100 ml of inoculum suspension before application. This spore suspension (containing approximately 3–4 × 10^4 spore/ml) is sprayed on the half-bloom inflorescence of the test plant raised in the pots or the experimental field. Approximately 30 ml of spray inoculum is required per inflorescence with a sprayer. The sprayed inflorescence is covered with a plastic bag to attain the required humidity for 30 h. Inflorescence sprayed with simple water can serve as a control. The inoculated inflorescence plants are placed in glass- and net-house at 22–25 °C, and note the disease severity, re-isolate the pathogen, and confirm Koch's postulate.

14

Molecular Techniques in the Identification of Fungal Cultures

In recent years several plant pathological laboratories have been using molecular techniques for identification of the fungal plant pathogens isolated from the diseased plants. These are specifically used for the confirmation of the pathogen identified based on their cultural and microscopic characteristics. With the help of fungal genomic studies, not only the genus but also the species of the pathogen is identified based on the genomic sequences available with NCBI, and provide the accurate identification of the fungal isolate. This identification method involves the following steps:

14.1. Isolation of Genomic DNA from the Fungal Pathogen

The fungal genomic DNA can be isolated by CTAB (Cetrimonium bromide) method given by Rogers and Benedich (1994).

Principle

The CTAB is a cationic detergent and serves two purposes. The detergent component present in the CTAB disrupts the cell membrane and nuclear membrane to expose the genetic components. After disruption, the proteinase in CTAB frees the DNA by destroying the histones. This takes place at 65 °C to optimize the enzyme activity of CTAB.

Materials required

1. Liquid nitrogen

   (It helps to break the cell wall and deactivates the enzyme activity).

2. PVP (polyvinyl-pyrrolidone), Mercapto-ethanol, Sodium meta-bisulphate

   (It helps to prevent phenolic oxidation).

   Note: Mercapto-ethanol should be opened in the fume hood only as it is volatile and toxic).

3. EDTA (Ethylene diamine tetra acetic acid)

   (It inactivates endonucleases by chelating the Mg^{2+}).

4. Chloroform and isoamyl alcohol

   (They are used to separate the proteins, polysaccharides, and pigments).

   Note: Always store this in amber-colored bottles as in the presence of light, chloroform is oxidized to highly poisonous gas phosgene. To prevent this, chloroform must be kept in dark brown colored bottles. Moreover, the bottles must be well closed to check the entry of air.

5. 70 and 100% ethanol

   (It helps to remove the salts and detergents).

6. CTAB buffer (2×), CTAB solution

   (It acts as a cationic detergent for cell lysis).

7. TE buffer (pH 8.0)—100 ml

   10 mM Tris base (pH 8.0); 1 mM EDTA (pH 8.0).

   (weigh out 0.12114 g of 10 mM Tris, dissolve it in a beaker in 40 ml of distilled water and adjust the pH to 8. Weigh out 0.03722 g of 1 mM EDTA, dissolve in another 40 ml of distilled water and adjust the pH to 8. Mix the two reagents and make the final volume to 100 ml using distilled water and autoclave the buffer).

8. Fungal mycelium

   Inoculate the sterilized potato-dextrose broth with the fungus under aseptic conditions and incubate it for the growth of fungal mycelium. Then collect the young growth of fungal mycelium and proceed for fungal DNA isolation.

Buffer preparation

1. 2% CTAB buffer (100 ml).

Content	Molecular weight (g)	Quantity required for 100 ml (g)
100 mM Tris (pH 8)	121.14	1.2114
20 mM EDTA (pH 8)	372.24	0.7448
1.4 M sodium chloride	58.44	8.1816
CTAB	364.45	2

2. 10% CTAB solution (100 ml).

Content	Molecular weight (g)	Quantity required for 100 ml (g)
10% CTAB	364.45	10
0.7 M sodium chloride	58.44	4.1

3. TE buffer (100 ml)

Content	Molecular weight (g)	Quantity required for 100 ml (g)
10 mM Tris (pH 8)	121.14	0.12114
1 mM EDTA (pH 8)	372.24	0.03722

4. Chloroform: isoamyl alcohol (24:1)

Equipment required

Centrifuge machine, hot water bath, vortex shaker, weighing machine, etc.

Procedure

1. Grind 100–150 mg of fungal mycelium in a pre-chilled mortar and pestle with a pinch of PVP and 50 μl of β-mercaptoethanol in the presence of liquid nitrogen.
2. Transfer the homogenized sample into an autoclaved 2 ml centrifuge tube and add 1 ml of pre-warmed extraction buffer.
3. Mix well and incubate the mixture at 65 °C for 30 min with occasional mixing by gentle inversion.
4. Add equal volumes of chloroform: isoamyl alcohol (24:1) and mix by inversion to emulsify. Centrifuge at 10,000rpm for 15 min at 4 °C.
5. After centrifugation, three layers will be formed.
   - Aqueous topmost layer—DNA.
   - Interphase—fine particles, cell debris.
   - Lower layer—chloroform.
6. Transfer the top aqueous layer to a clean tube and add 1/10th volume of 10% CTAB solution and an equal volume of chloroform: isoamyl alcohol (24:1) and mix gently by inversion.
7. Centrifuge at 10,000 rpm for 15 min at 4 °C.
8. Transfer the aqueous layer to a clean tube, add 0.6 volumes of chilled isopropanol and mix by quick gentle inversion till the DNA precipitates. Keep at −20 °C for half an hour for complete precipitation.

9. Centrifuge at 10,000 rpm for 15 min at 4° C. Gently pour off the supernatant.
10. Wash the DNA pellet with 70% ethanol followed by 100% ethanol.
11. Centrifuge for 3 min at 10,000 rpm and decant ethanol.
12. Air dry the pellet, dissolve it in 50 µl of TE buffer or sterile water, and store it at −20 °C.

The genomic DNA of fungus can be isolated by this method and the sample can be stored at −20° C for future use.

14.2. Quantification of DNA by Nanodrop Spectrophotometer

Aim

To quantify the DNA present in a given sample and to check the purity of DNA.

Principle

The nanodrop spectrophotometer is designed for measuring nucleic acid concentrations in sample volumes of one microliter. Purines and pyrimidines in the nucleic acids show absorption maxima around 260 nm if the DNA sample is pure and is without any protein or RNA contaminations, whereas proteins show peak absorbance at 280 nm. Absorbance is recorded at both wavelengths and the purity is indicated by $A_{260/280}$. If the value of $A_{260/280}$ lies between 1.8 and 2.0, the DNA is pure and is contaminated with RNA if the value is above 2.0, and if below 1.8, the sample is contaminated with proteins.

Materials required

Tissue paper, wash bottle with distilled water, sample, blank, and pipette.

Procedure

1. The quantity of DNA is checked using the "JENWAY Genova Nano" nanodrop spectrophotometer.
2. Open the sampling arm and pipette 1 µl of distilled water onto the lower measurement pedestal.
3. Close the sampling arm and set the reading to zero with distilled water which serves as the blank.
4. Open the sampling arm and wipe the distilled water using blotting paper.
5. Pipette 1 µl of the sample to the sampling arm and close the sampling arm.

6. Read the $A_{260/280}$ value and the quantity of the DNA per µl of the sample.
7. When the measurement is complete, wipe the sampling arm with blotting paper.

The concentration of the DNA can be checked by using a nanodrop spectrophotometer. If the value obtained is between 1.8 and 2.0, the DNA is pure and is contaminated with RNA if the value is above 2 and is below 1.8, the sample is contaminated with proteins.

14.3. PCR (Polymerase Chain Reaction) Amplification of Isolated DNA

Aim

To amplify the ITS-rDNA region of the isolated fungal DNA using fungal-specific primers (ITS 1 and ITS 4).

Principle

Polymerase chain reaction (PCR) involves the primer-mediated enzymatic amplification of the DNA. PCR is based on using the ability of DNA polymerase to synthesize new strands of DNA complementary to the offered template strand. The PCR reaction mixture contains the template DNA, primers, dNTPs, buffers, Mg and Taq polymerase. Primers are the oligonucleotide or short stretch of single-stranded DNA that binds to the template DNA and provides a free 3'OH group to which the Taq polymerase can add dNTPs; thereby synthesizing the complementary strands of both template strands.

Materials required

The PCR reaction requires the following components:

1. DNA template: the double-stranded DNA (dsDNA) of interest is separated from the sample.
2. DNA polymerase: this enzyme does not break down at high temperatures. Taq polymerase attaches nucleotides to a DNA template, thereby copying the DNA.
3. Oligonucleotide primers: short pieces of single-stranded DNA (often 20–30 bp) which is complimentary to the 3′ ends of the sense and antisense strands of the target sequence.
4. Deoxynucleotide triphosphates (dNTPs): single units of bases A, T, G, and C (dATP, dTTP, dGTP, and dCTP) provide the energy for polymerization and the building blocks for DNA synthesis.

5. Buffer system: Includes magnesium and potassium to provide the optimal conditions for DNA denaturation and renaturation, also important for polymerase activity, stability, and fidelity.
6. PCR tubes.
7. Pipettes.

Equipment required

PCR machine

Steps involved in PCR

1. Denaturation

   In this step, the reaction mixture is heated to a temperature of 92–94 °C. During this, the double-stranded DNA is denatured to single strands due to the breakage of weak hydrogen bonds.

2. Annealing

   The temperature used is 54–60°C. In this step, the primers bind to their complementary sequence in the template DNA.

3. Elongation

   It is also called an extension and usually occurs at a temperature of 72–80 °C. During this the polymerase sequentially adds bases to the 3' primer, extending the DNA sequence in the 5' to 3' direction.

Procedure

1. 25 µl of reaction mixture is added to a PCR tube. PCR tubes are special thin-walled tubes of uniform thickness which are used to ensure rapid and equal temperature changes throughout the reaction volume.
2. The PCR reaction mixture contains (each tube).

Takara master mix	12.5 µl
Reverse primer	0.5 µl
Forward primer	0.5 µl
Template DNA	1 µl
dH_2O	10.5 µl

3. The thermocycler is programmed for desired times and temperatures of denaturation, annealing, and elongation.

4. The amplification is carried out with the following program.

   LID: 98 °C

SI. No.	Steps	Temperature (°C)	Time (min)
1	Initial denaturation	95	5.00
2	Denaturation	95	0.45
3	Primer annealing	51.2	0.30
4	Primer extension	72	1.00
Go to step 2 and repeat up to 30 cycles			
5	Final extension	72	10.00
6	Final hold	4	10.00
7	End		

5. A negative control is maintained without adding template DNA to the reaction mixture.
6. At the end of the PCR run, the PCR tubes are taken out of the thermocycler. View the PCR product by electrophoresis which is carried out in 1% agarose gel stained with Ethidium bromide and view it under UV light.

The size of the PCR products is visible.

Gradient polymerase chain reaction

Aim

To determine the optimal annealing temperature.

Principle

PCR involves the primer-mediated enzymatic amplification of DNA. PCR is based on using the ability of DNA polymerase to synthesize new strands of DNA complementary to the offered template strand. The PCR reaction mixture contains the template DNA, primers, dNTPs, buffers, Mg^{2+}, and Taq polymerase. Primers are the oligonucleotide or short stretch of single-stranded DNA that binds to the template DNA and provides a free 3'OH group to which the Taq polymerase can add dNTPs; thereby synthesizing the complimentary strands of both template strands.

Materials required

The PCR reaction requires the following components:

1. DNA template: the double-stranded DNA (dsDNA) of interest is separated from the sample.
2. DNA polymerase: this enzyme does not break down at high temperatures. Taq polymerase attaches nucleotides to a DNA template, thereby copying the DNA.
3. Oligonucleotide primers: short pieces of single-stranded DNA (often 20–30 bp) which is complimentary to the 3' ends of the sense and anti-sense strands of a target sequence.
4. Deoxynucleotide triphosphates (dNTPs): single units of bases A, T, G, and C (dATP, dTTP, dGTP, and dCTP) provide the energy for polymerization and the building blocks for DNA synthesis.
5. Buffer system: includes magnesium and potassium to provide the optimal conditions for DNA denaturation and renaturation, also important for polymerase activity, stability, and fidelity.
6. PCR tubes
7. Pipettes

Procedure

1. 25 μl of the reaction mixture is added to each PCR tube. The PCR reaction mixture includes:

Takara master mix	12.5 μl
Reverse primer	0.5 μl
Forward primer	0.5 μl
Template DNA	1 μl
dH_2O	10.5 μl

2. The thermocycler is programmed with desirable time and temperature for denaturation and elongation while a range of temperatures is used for annealing at desired times.

The PCR amplification is carried out with the following program:

SI. No.	Steps	Temperature (°C)	Time (min)
1	Initial denaturation	94	2.00
2	Denaturation	94	0.45
3	Primer annealing	50–62	1.00
4	Primer extension	72	2.00
Go to step 2 and repeat up to 29 cycles			
5	Final extension	72	10.00
6	Final hold	4	10.00
7	End		

3. A negative control is maintained without adding template DNA to the reaction mixture.
4. At the end of the run the tubes are taken out and electrophoresis is carried out in 1% agarose gel stained with ethidium bromide and viewed under UV light.

Observations

The result of gradient PCR should be viewed under a UV transilluminator to find out the best amplification among various temperatures. The size of the PCR product should also be observed (Fig. 14.1).

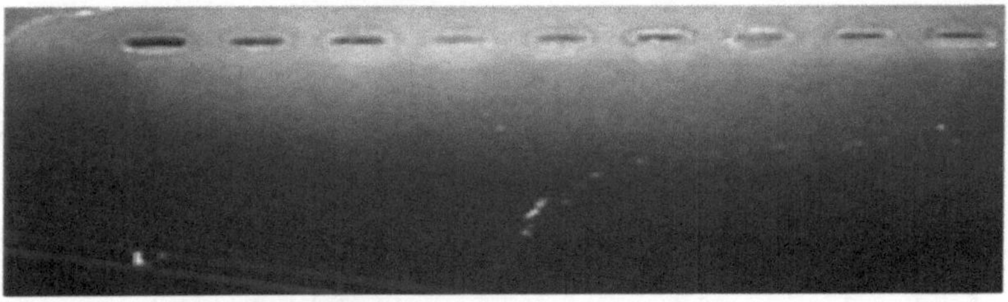

Fig. 14.1: Gel profile of gradient PCR product.

14.4. Agarose Gel Electrophoresis

Aim

To separate, identify and purify DNA fragments and to visualize it.

Principle

The negatively charged DNA molecules migrate toward the positive charge under the influence of constant current and the separation depends on the mass and charge of DNA. The DNA molecules are forced to move through the agarose gel pores. The location of the DNA in the gel is determined by staining with a low concentration of intercalating fluorescent ethidium bromide dye under UV light.

The speed of movement of the DNA in the gel depends on:

1. Molecular weight of the DNA

   The higher the molecular weight of DNA, the slower will be its movement through the gel.

2. Concentration of agarose

   Agarose serves as a sieve for the separation of macromolecules. The greater the concentration of agarose, the smaller the pore size. The usual concentration of agarose use ranges from 0.7 to 2%.

3. Confirmation of DNA

   A circular DNA molecule moves faster than a linear molecule on the gel because a linear molecule exerts greater friction than a circular molecule on the gel.

4. Strength of electric field applied

   The electric field applied during agarose gel electrophoresis ranges from 70 to 100 V. An increase above this range results in faster movement of the DNA but the process generates a greater amount of heat.

5. Composition and ionic strength of electrophoresis buffer

   In the absence of ions, electric conductivity is minimal and DNA migrates very slowly.

Materials required

- DNA sample
- 0.8% agarose

Agarose is a polysaccharide extracted from the seaweed. The long chain of (polymer) agarobiose creates the agarose sugar. Agaropectin is removed from the agar to form agarose. By forming the hydrogen bond between adjacent molecules, it creates a three-dimensional matrix that forms pores. The melting temperature of the agarose is near the boiling temperature (~95 °C) whereas the gelling temperature is ~37–43 °C.

- 1X TAE buffer

  2 ml of 50X TAE buffer + 98 ml of water for 100 ml of 1X TAE buffer.

  Preparation of 50X TAE buffer

  Dissolve 242 g of TRIS base in water, add 57.1 ml glacial acetic acid and 100 ml of 500 mM EDTA (pH 8.0) solution, and bring the final volume up to 1 L.

- 6X gel loading dye

  Contains 0.25% bromophenol blue, 0.25% xylene cyanol, and 30% glycerol.

- Ethidium bromide

  It is a fluorescent dye used in tagging DNA. DNA does not have its color, EtBr intercalates with the DNA bases and gives a fluorescent color under UV light.

- Molecular weight marker

  1 kb ladder–1000 bp to 10,000 bp.

  100 bp ladder–100 to 1000 bp.

Equipment

Gel electrophoresis unit.

Procedure

1. Prepare 100 ml of 1X TAE buffer in a conical flask.
2. Weigh the required quantity (0.8 g for DNA, 1 g for PCR product) of agarose and add to 1X TAE buffer in a conical flask.

3. Melt the agarose and dissolve it by heating.
4. Cool the solution.
5. Add 2 µl of EtBr to the solution. Mix carefully taking care to avoid the formulation of bubbles.
6. Wipe the gel casting tray comb with alcohol and fix it properly in the casting apparatus.
7. Pour the melted gel into the casting tray and allow it to solidify at room temperature.
8. After solidification, pull out the comb to make the well in the gel.
9. Place the gel in the buffer tank containing 1X TAE buffer.
10. Take 3 µl of DNA and mix it well with 3 µl of gel-loading dye. Bromophenol blue present in the dye moves ahead of the DNA sample, so it becomes easy to track the movement of the DNA on the gel. Glycerol increases the viscosity of the DNA sample and therefore helps the DNA to settle in the well.
11. Carefully load the 6 µl sample into the well using a micropipette. Load a molecular weight ladder in one lane. DNA should be loaded only at the negative terminal of the tank.
12. Switch on the current and allow the DNA to run on the gel. When the dye reaches 2/3rd of the gel, switch off the current and view the gel under a UV light transilluminator.

Observations

Presence of intact DNA bands.

Precautions

- Do not get exposed to UV light.
- Load DNA at the negative terminal only.
- Discard EtBr-stained agarose gel carefully, by adding sufficient water to reduce the concentration of EtBr to 0.5 µg/ml. Add one volume of 0.5M $KMnO_4$, mix carefully, and then add one volume of 2.5 N HCl. Mix carefully and allow the solution to stand at room temperature for several hours. Add one volume of 2.5 N NaOH. Mix carefully and then discard the solution (Fig. 14.2).

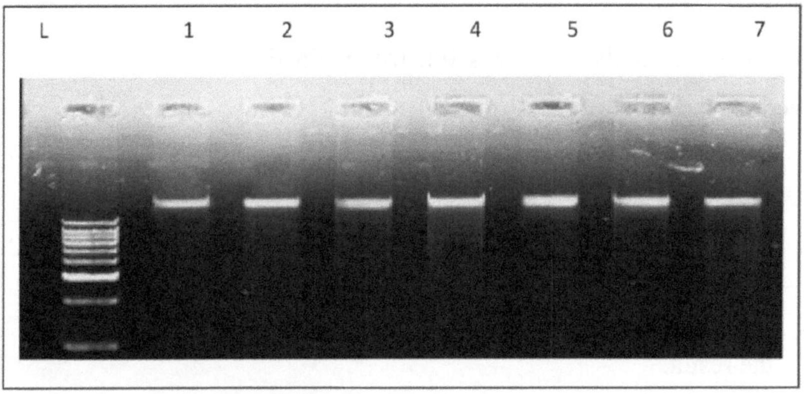

Fig. 14.2: Gel profile of PCR product.

14.5. BLASTn (Nucleotide Query with Nucleotide Database) for Data Interpretation and Identification of Fungal Isolate

After obtaining the gene sequence after out sourcing the PCR product, the result obtained should be interpreted by using *in-silico* analysis. The National Center for Biotechnology Information (NCBI) is part of the United States National Library of Medicine (NLM), a branch of the National Institute of Health (NIH). The NCBI has the responsibility for making available the GenBank DNA sequence database since 1992. The Basic Local Alignment Search Tool program was developed by Stephen Altschul of NCBI in 1990 and has become one of the most popular programs for sequence analysis. BLAST uses heuristics to align a query sequence with all sequences in a database. The objective is to find high-scoring un-gapped segments among related sequences. The existence of such segments above a given threshold indicates pairwise similarity beyond random chance, which helps to discriminate related sequences from unrelated sequences in a database. BLAST is a family of programs that includes BLASTN, BLASTP, BLASTX, TBLASTN, and TBLASTX.

BLASTN is widely used for fungal sequence analysis as it has DNA as genetic material. The BLAST output provides a list of pairwise sequence matches ranked by statistical significance. The significance score helps to distinguish evolutionary-related sequences from unrelated ones. Generally, only hits above a certain threshold are displayed.

Aim

To do BLASTN analysis of the sequence obtained.

Procedure

1. Go to the NCBI homepage and save the nucleotide sequence in FASTA format.
2. Go to the BLAST homepage from the NCBI website and select BLASTN. Upload the query sequence.
3. After entering the query sequence, click the "BLAST" option to obtain the result.
4. Interpret the result obtained (Figs. 14.3–14.5).

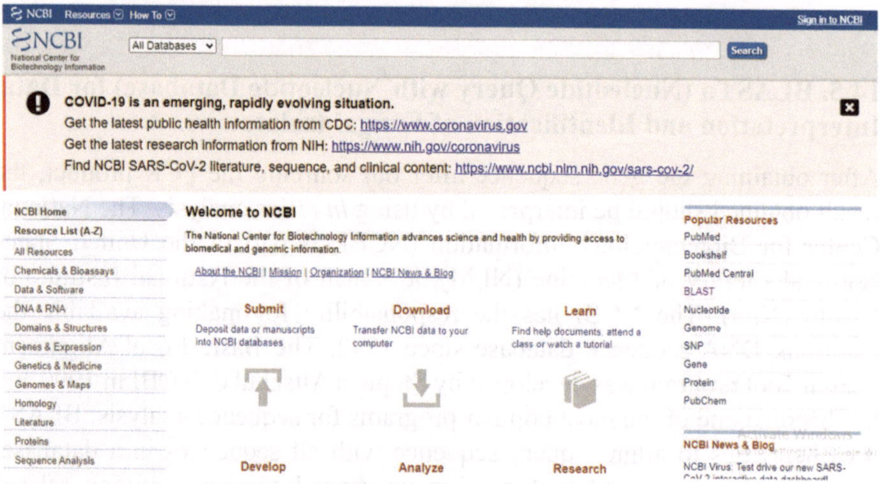

Fig. 14.3: Homepage of NCBI.

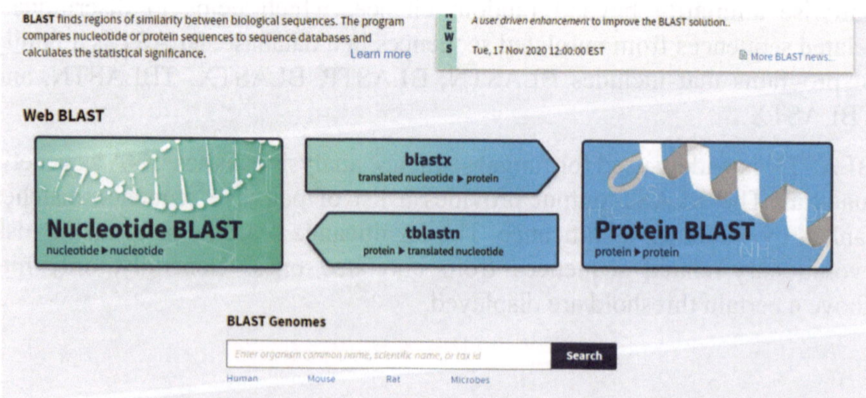

Fig. 14.4: Select Nucleotide BLAST to analyze fungal DNA sequence.

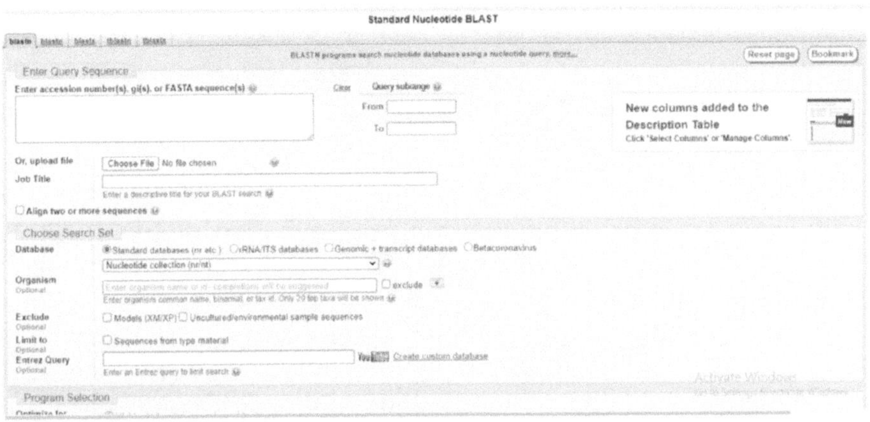

Fig. 14.5: Upload the nucleotide sequence in FASTA format and blast it to get the results.

Observations

The sequence we uploaded is compared with the nearest match obtained to identify the fungi by selecting it as a similarity sequence from obtained results.

Appendices

1. Units of Measurements

1.1 Decimal fractions

SI. no.	Fraction	Symbol	Prefix	Abbreviations
1	Thousandth	10^{-3}	Milli	M
2	Millionth	10^{-6}	Micro	μ
3	Billionth	10^{-9}	Millimicro (or nano)	mμ (mμ)
4	Trillionth	10^{-12}	Micromicro (or pico)	μμ (μμ)

1.2 Molar concentration

SI. no.	Mole fraction	Name	Abbreviations	Amount/l	Amount/ml
1		Molar	Mol	1 mol/l	1 m mol/ml
2	10^{-3}	Millimolar	mM	1 millimol/l	1 μ mol/ml
3	10^{-6}	Micromolar	μM	1 μmol/l	1 mμ mol/ml
4	10^{-9}	Nanomolar	nM	1 mμ mol/l	1 μμ mol/ml

1.3 Weights

SI. no.	Name	Abbreviations	Fraction of gram (g)
1	Gram	g	
2	Milligram	mg	10^{-3}
3	Microgram (gamma)	μg (γ)	10^{-6}
4	Nanogram	ng	10^{-9}
5	Picogram	pg	10^{-12}

1.4 Volume

SI. no.	Name	Abbreviations	A fraction of a liter
1	Liter	1	
2	Milliliter	ml	10^{-3}
3	Microliter (lambda)	μl (γ)	10^{-6}

1.5 Length

Sl. no.	Name	Abbreviations	Fraction of a meter
1	Centimeter	cm	10^{-2}
2	Millimeter	mm	10^{-3}
3	Micron	μ	10^{-6}
4	Millimicron	mμ	10^{-9}
5	Angstrom unit	Å	10^{-10}

2. Standard Solutions (1N)

2.1	Acetic acid	Dilute 57.1 ml of 17.5 N acid to 1 l with water.
2.2	Formic acid	Dilute 46 ml to 1 l with water.
2.3	Hydrochloric acid	Dilute 82.36 ml of 36.8–38.0% acid to 1 l with water.
2.4	Nitric acid	Dilute 63.0 ml of 69–71% acid to 1 l with water.
2.5	Perchloric acid	Dilute 84.0 ml of 72% acid to 1 l with acetic acid.
2.6	Sulfuric acid	Dilute 27.7 ml of 95–98% acid to 1 l with water.
2.7	Ammonium hydroxide	Dilute 66.6 ml of 15 N solution to 1 l with water.
2.8	Potassium hydroxide	56.1 g in 1 l of water.
2.9	Sodium hydroxide	41.24 g of 97% NaOH in 1 l of water.
2.10	Sodium bicarbonate	84.01 g in 1 l of water.
2.11	Iodine	200 g iodate-free potassium iodide in 300–400 ml of water and add 127 g of sublimed iodine; make up the volume to 1 l.
2.12	Sodium thiosulphate	250 g of $Na_2S_2O_3 \cdot 5H_2O$ in boiled distilled water; make up the volume to 1 l with water and add 1 g of anhydrous Na_2CO_3 to stabilize the solution.
2.13	Potassium permanganate	32.5 g in 1 l of water.

Standardize the solutions before use.

3. Indicator Solutions

3.1	Bromo-phenol blue	Dissolve 0.1 g in 7.5 ml of 0.02 M NaOH and make up the volume to 250 ml with water. 0.1% in water may also be used. Its color is yellow in acid and blue in alkaline solutions.
3.2	Bromo-cresol green	Dissolve 0.1 g in 7.25 ml of 0.02 M NaOH and make up the volume to 250 ml with water. Its color is yellow in acid and blue in alkaline solutions.

3.3	Bromo-cresol purple	Dissolve 0.1 g in 9.25 ml of 0.02 M NaOH and make up the volume to 250 ml with water. Its color is yellow in acid and purple in alkaline solutions.
3.4	Bromo-thymol blue	Dissolve 0.1 g in 8 ml of 0.02 M NaOH and make up the volume to 250 ml with water.
3.5	Methyl orange	Dissolve 0.025 g of free acid in 250 ml of water or dissolve 0.025 g of sodium salt in 3.8 ml of 0.1 N HCl and make up the volume to 250 ml with water. Its color is red in acid and yellow in alkaline solutions.
3.6	Methyl red	Dissolve 0.1 g in 18.6 ml of 0.02 M NaOH and make up the volume to 250 ml with water. 0.5% in 95% ethanol may also be used. Its color is red in acid and yellow in alkaline solutions.
3.7	Phenolphthalein	Dissolve 0.25 g in 250 ml of 95% ethanol. It is colorless in acid and pink in alkaline solutions.

4. Stains

4.1 Cotton blue lactophenol

Cotton blue (1% aqueous solution)	5 ml
Phenol crystals	20 g
Lactic acid	20 g
Glycerin	40 ml

4.2Lugol's solution

Iodine	1 g
Potassium iodide	2 g
Distilled water	300 ml

5. Disinfectants

5.1 Mercuric chloride solutions

Stock solution

Mercuric chloride	20 g
Concentrated HCl	100 ml

Dilute 5 ml of the stock solution to 1 l with water.

5.2 Sodium hypochlorite

Make fresh solutions (1%) of sodium hypochlorite (NaOCl). Acidic solution (pH 4) is more efficient than commercial bleach solutions which are alkaline. Deteriorates faster during storage. Alternatively, use 0.35% (w/v) aqueous solution of commercial bleaching powder ($Ca(OCl)_2$).

6. Plant Nutrient Solutions (Hoagland solution) (for *in vitro* Plant-Cultures)

Solution 1	**g/l**
KH_2PO_4	0.136
KNO_3	1.02
$Ca(NO_3)_2$	0.492
$MgSO_4 \cdot 7H_2O$	0.49

Solution 2	**g/l**
$NH_4H_2PO_4$	0.23
KNO_3	1.02
$Ca(NO_3)_2$	0.492
$MgSO_4 \cdot 7H_2O$	0.49

Solution A	**mg/l**
H_3BO_3	2.86
$MnCl_2 \cdot 4H_2O$	1.81
$ZnSO_4 \cdot 7H_2O$	0.22
$CuSO_4 \cdot 5H_2O$	0.08
$H_2MoO_4 \cdot H_2O$	0.09

Solution B

Dissolve 26.1 g ethylenediamine tetra-acetic acid in 268 ml of 1 N KOH. Then add 24.9 g $FeSO_4.7H_2O$ and dilute to 1 l. Aerate overnight to produce a stable ferric complex. The pH of the solution should be about 5.5.

Add 1 ml of **solution A** to either 1 l of **solution 1 or 2**. Add 1 ml of **solution B** to it. Adjust the pH to 6 with 0.1 N H_2SO_4.

Index